Oscillator Circuits & Projects

Written By
Stephen Kamichik

PUBLICATIONS

A Division of Howard W. Sams & Company
A Bell Atlantic Company
Indianapolis, IN

International Standard Book Number: 0-7906-1111-2
Library of Congress Card Catalog Number: 97-65789

Acquisitions Editor: Candace M. Hall
Editors: Natalie F. Harris
Assistant Editors: Pat Brady, Loretta L. Leisure
Typesetting: Natalie Harris
Cover Design: Christy Pierce
Graphics Conversion: Natalie Harris, Walt Striker, Terry Varvel
Illustrations: Stephen Kamichik, Bob Medaris, Bill Skinner

PRINTED IN THE UNITED STATES OF AMERICA

9 8 7 6 5 4 3 2 1

Contents

Contents

Preface

Oscillator circuits were not formally taught at the universities where I studied for my technology and engineering degrees. Very few books have been written exclusively about oscillator circuits. Electronics is a vast field; therefore, some areas of expertise must be acquired by the practicing technician and engineer. This book was written to be used as a textbook and project book on oscillator circuits.

This book may be used by hobbyists, technicians and engineers. If you have no knowledge of electronics, I suggest that you read *Semiconductor Essentials*, written by the author. Everybody can build and enjoy the projects described at the end of this book.

This book is divided into two parts. Part 1 consists of ten chapters, and discusses oscillator circuits. Part 2 consists of eight chapters, and describes oscillator projects.

Chapter 1 is an introduction to oscillator circuits. Chapter 2 discusses negative and positive feedback. Chapter 3 describes several sinusoidal oscillator circuits. Chapter 4 shows you how to design sinusoidal oscillators. Chapter 5 is about the Schmitt trigger. Chapter 6 discusses transistor multivibrator circuits, and Chapter 7 discusses integrated circuit multivibrator circuits. Chapter 8 covers blocking oscillators. Chapter 9 is about negative-resistance devices and oscillators. Chapter 10 discusses linear sawtooth waveform generators. There are several problems at the end of each chapter of Part 1. The problem solutions are at the end of the book in the appendix.

Chapter 11 gives a step-by-step discussion on how to build a function generator. Chapter 12 details the construction of an auto-ranging digital capacitance meter. Chapter 13 describes the construction of a 30 MHz frequency counter. Chapter 14 details the construction of a solar powered generator. Chapter 15 describes the construction of an electronic siren. Chapter 16 gives the construction details of two electronic organs. Chapter 17 describes the construction of two radio frequency tone transmitters. Chapter 18 details the construction of two FM voice transmitters.

Part 1

Oscillator Circuits

Chapter 1

Introduction

An amplifier is a circuit that produces an amplified version of its input signal. Most amplifier circuits have negative feedback; that is, a portion of the output signal is fed back to the input circuit, out of phase with the input signal. *Figure 1-1* is a block diagram of an amplifier circuit.

An oscillator is a circuit that produces a waveform of a definite shape and frequency. No input signal is required. A portion of the output signal is fed to the input circuit, in phase with the starting power. This is known as positive or regenerative feedback. *Figure 1-2* is a block diagram of an oscillator circuit.

Different types of oscillators are used to produce different types of waveforms. Oscillators can be used to generate sinewaves, squarewaves, sawtooth waveforms, etc.

Feedback is accomplished by inductive, capacitive or resistive coupling between the output and the input of a circuit. A variety of circuits can be used to produce feedback of the proper phase and amount. Oscillators may be designed to generate waveforms, from low audio frequencies to very high radio frequencies, by using components of the proper values.

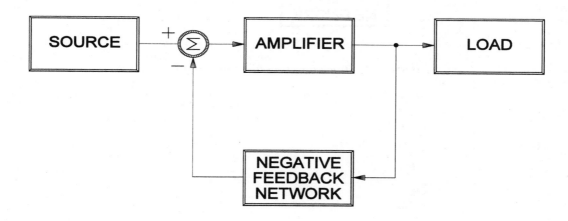

Figure 1-1. Block diagram of an amplifier.

Most oscillators are classified as either tuned inductive-capacitance (LC) types or relaxation resistance-capacitance (RC) types. At frequencies below 100 kHz, the required inductors of an LC oscillator are large with non-ideal characteristics.

Electromechanical oscillators are used to generate waveforms whose frequencies are very stable, because a mechanical device with a constant source of vibration is used as a frequency reference. A crystal oscillator is an electromechanical oscillator.

There are oscillatory conditions in nature. When a wind of sufficient force hits a structure (such as a bridge) of a certain resonant frequency, it can go into oscillation and self-destruct. Once the structure starts vibrating at its resonant frequency, its mass reinforces and increases the amplitude of the vibration until it self-destructs. Bridges and buildings must be designed with very high resonant frequencies in mind.

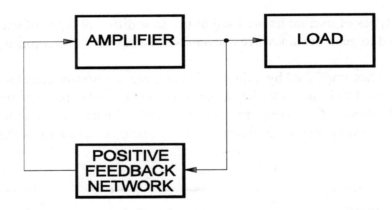

Figure 1-2. Block diagram of an oscillator.

Problems

Problem 1-1.   Compare amplifiers and oscillators.

Problem 1-2.   Why are LC oscillators not used to generate waveforms whose frequencies are less than 100 kHz?

Problem 1-3.   What is an electromechanical oscillator?

Problem 1-4.   Name a type of electromechanical oscillator.

Chapter 2

Feedback

Harold Black invented the negative feedback amplifier in 1928. He was working as an electronics engineer for the Western Electric company. Negative feedback was first used to design amplifiers with stable gain for use in telephone repeaters. Today, most electronic circuits incorporate some form of feedback. Feedback is also used in disciplines outside of electronics, such as in the modeling of biological systems.

Feedback is either negative (degenerative) or positive (regenerative). Negative feedback is used in amplifier designs and positive feedback is used in oscillator designs. Negative feedback tends to stabilize a circuit. Positive feedback usually destabilizes a circuit and is therefore ideal for designing oscillators. Positive feedback *can* stabilize a circuit, though, which is useful in the design of active filters.

Negative feedback has many desirable effects on an amplifier circuit. Negative feedback causes the gain of the amplifier to be less sensitive to component value changes. This can be useful in circuits that are subject to temperature variations. Most satellite circuits employ negative feedback.

Negative feedback reduces nonlinear distortion. The output signal is proportional to the input signal, but independent of the input signal level.

Negative feedback is used in automatic gain control circuits. It also reduces the effects of noise. Noise is any unwanted electrical signal generated by circuit components and extraneous interference.

Negative feedback can be used to control the input and output impedances of an amplifier. Selecting the appropriate feedback topology can increase or decrease the input and output impedances as required. Negative feedback also extends the bandwidth of the amplifier.

The desirable features of negative feedback are obtained at the expense of a reduction in the gain of the amplifier. The gain reduction factor is called the *amount of feedback*. It is the factor by which the amplifier is desensitized.

Figure 2-1. Block diagram of an amplifier without feedback.

Feedback Structure

The gain of an amplifier without negative feedback is its open-loop gain, A. *Figure 2-1* is a block diagram of an amplifier without feedback. If a feedback loop is added to the amplifier, as shown in *Figure 2-2*, some of the output voltage, B, is fed back to the input. The input signal is now the original signal plus the feedback voltage. The term B is called the feedback factor. Denoting A' as the amplifier gain with feedback (closed-loop gain), and AB as the loop gain, the basic feedback equation is A' = A/(1 + AB).

The quantity AB is the loop gain of the amplifier. The loop gain is positive when the feedback is negative. The quantity (1 + AB) is the amount of feedback.

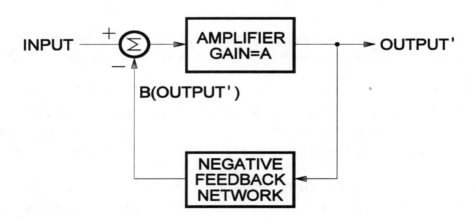

Figure 2-2. Block diagram of an amplifier with negative feedback.

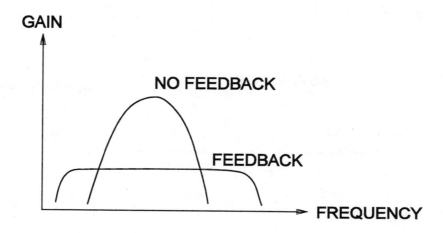

Figure 2-3. *Negative feedback and the response of an amplifier.*

When loop gain AB is very large, that is, AB >> 1, then A' = 1/B. This implies that the gain of the feedback amplifier is determined by the feedback network. If a large amount of negative feedback is used, the feedback and input signals become identical. This is how the tracking of the two input terminals of an operational amplifier functions.

Properties of Negative Feedback

Negative feedback desensitizes the closed-loop gain of an amplifier. This property may be written as dA'/A' = dA/A(1 + AB). This states that the percentage change in A' (due to variations in some circuit parameter) is smaller than the change in A by the amount of negative feedback. Therefore, the amount of feedback, (1 + AB), is also called the desensitivity factor.

Since negative feedback increases the bandwidth of an amplifier, the new midband gain of the amplifier is Am/(1 + AmB), where Am is the open-loop midband gain. The new upper 3dB frequency is:

$$w_{HF} = w_H(1 + AmB)$$

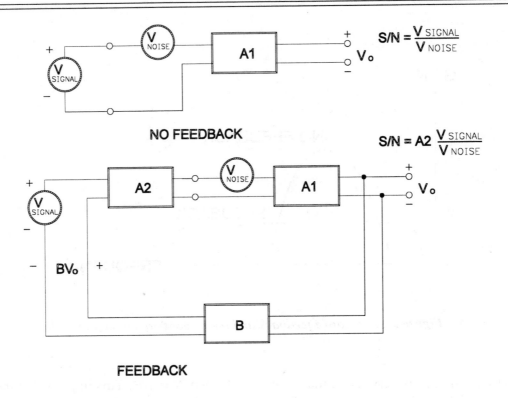

NO FEEDBACK

$$S/N = \frac{V_{SIGNAL}}{V_{NOISE}}$$

$$S/N = A2 \frac{V_{SIGNAL}}{V_{NOISE}}$$

FEEDBACK

Figure 2-4. Negative feedback improves signal-to-noise ratio.

This means that the upper 3dB frequency is increased by the amount of feedback. The new lower 3dB frequency is $w_{LF} = w_L/(1 + AmB)$, which means that the lower 3dB frequency is lowered by the amount of feedback. The effect of negative feedback on the frequency response of an amplifier is shown in *Figure 2-3*.

Negative feedback reduces the noise in an amplifier, that is, it increases the ratio of signal to noise. The improvement in the signal-to-noise ratio by negative feedback is only possible if the noisy stage is preceded by a relatively noise-free stage, as shown in *Figure 2-4*.

The amplifier transfer characteristic is linearized by negative feedback. This is shown in *Figure 2-5*. Negative feedback makes the closed-loop gain of an amplifier relatively independent of its open-loop gain, at the price of a reduction in gain.

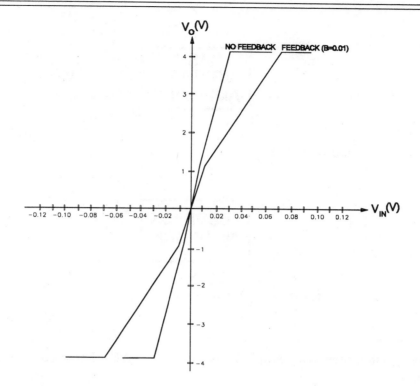

Figure 2-5. *Negative feedback reduces nonlinear distortion.*

Negative Feedback Topologies

There are four negative feedback topologies: *series-shunt* or voltage-sampling series-mixing; *shunt-series* or current-sampling shunt-mixing; *series-series* or current-sampling series-mixing; and *shunt-shunt* or voltage-sampling shunt-mixing.

Series-Shunt Feedback

Series-shunt feedback is also known as voltage-sampling series-mixing feedback, shown in *Figure 2-6*. The feedback network samples the output voltage. The feedback signal is a voltage that is mixed in series with the source voltage. The series-shunt topology stabilizes the voltage gain, increases the input resistance, and decreases the output resistance of the circuit. The non-inverting operational-amplifier circuit of *Figure 2-7* is a series-shunt feedback circuit.

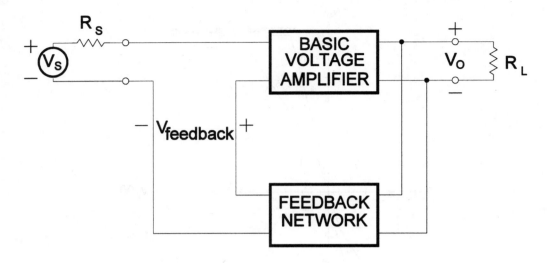

Figure 2-6. Series-shunt topology.

The gain of a series-shunt feedback amplifier is $A' = A/(1 + AB)$. The input resistance is $R'_{IN} = R_{IN}(1 + AB)$.

The output resistance is $R'_O = R_O/(1 + AB)$.

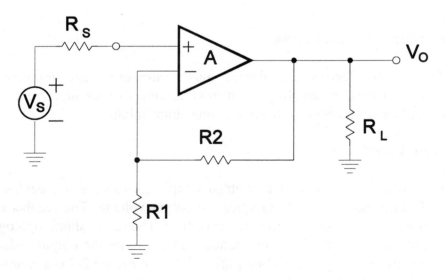

Figure 2-7. Series-shunt feedback circuit.

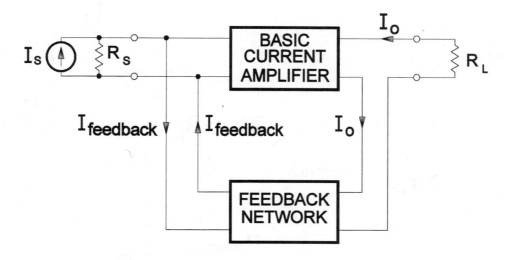

Figure 2-8. *Shunt-series topology.*

The practical series-shunt feedback amplifier is not an ideal voltage-controlled voltage source because the feedback network is usually passive and therefore loads the basic amplifier. The voltage amplifier is usually a series-shunt circuit.

Shunt-Series Feedback

The shunt-series topology is also called the current-sampling shunt-mixing topology, shown in *Figure 2-8*. The feedback network samples the output current and feeds it in shunt with the source current. The shunt-series topology stabilizes the current gain, reduces the input resistance and increases the output resistance of the circuit. A shunt-series feedback amplifier is shown in *Figure 2-9*. The current being sampled is not the output current but rather the emitter current of Q2. This is done for design convenience. The output current and the Q2 emitter current are approximately equal.

The gain of a shunt-series amplifier is $A' = A/(1 + AB)$. The input resistance is $R'_{IN} = R_{IN}/(A + B)$ and the output resistance is:

$$R'_o = R_o(1 + AB).$$

The current amplifier is usually the basis of a shunt-series circuit.

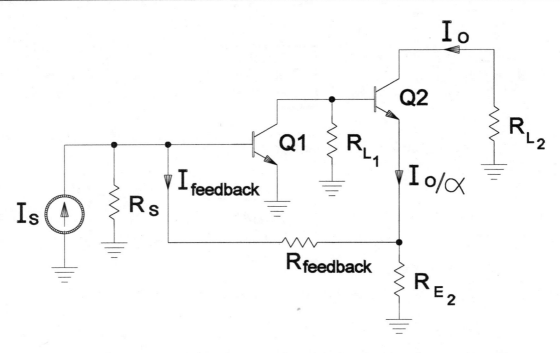

Figure 2-9. Shunt-series feedback circuit.

Series-Series Feedback

Series-series feedback is also known as current-sampling series-mixing feedback, shown in *Figure 2-10*. The transconductance amplifier is usually the basis of a series-series feedback circuit because its input signal is voltage and its output signal is current.

The gain of a series-series circuit is $A' = A/(1 + AB)$. The input and output resistances are increased: $R'_{IN} = R_{IN}(1 + AB)$ and $R'_O = R_O(1 + AB)$.

A series-series feedback circuit is shown in *Figure 2-11*. The current sampled is the emitter current of Q3. This is done for design convenience because the output current and the emitter current of Q3 are approximately equal.

Shunt-Shunt Feedback

Shunt-shunt feedback is also called voltage-sampling shunt-mixing feedback, shown in *Figure 2-12*. The transresistance amplifier is usually the basis of a shunt-shunt feedback circuit because the input signal is current and the output signal is voltage.

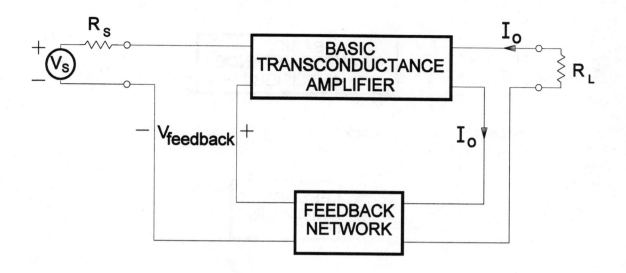

Figure 2-10. *Series-series topology.*

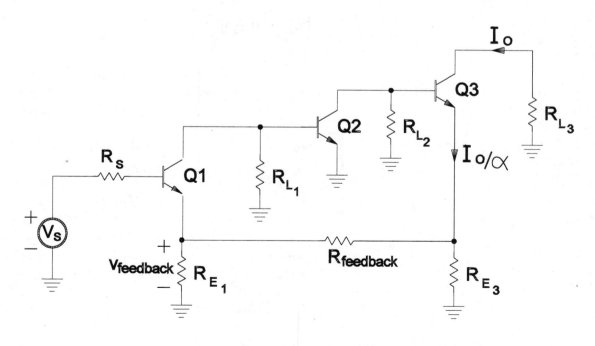

Figure 2-11. *Series-series feedback circuit.*

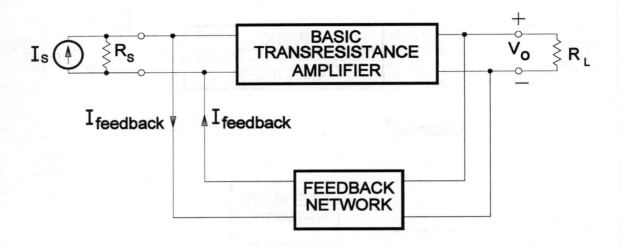

Figure 2-12. *Shunt-shunt topology.*

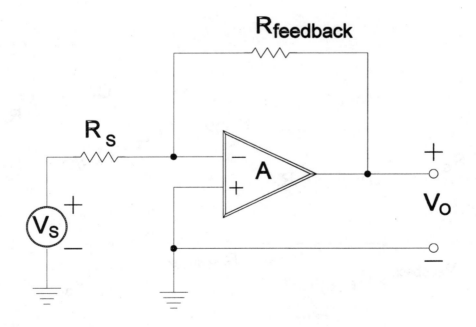

Figure 2-13. *Shunt-shunt feedback circuit.*

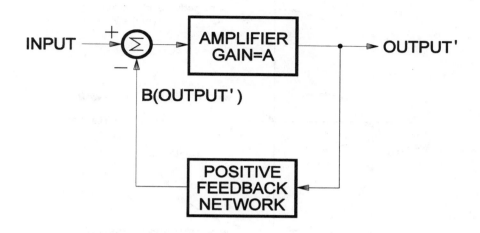

Figure 2-14. *Block diagram of a positive-feedback loop.*

The gain of a shunt-shunt circuit is A' = A/(1 + AB). The input and output resistances of the circuit are decreased:

$$R'_{IN} = R_{IN}/(1 + AB) \text{ and } R'_{O} = R_{O}/(1 + AB).$$

A shunt-shunt circuit is shown in *Figure 2-13*.

Positive Feedback

The structure of a sinusoidal oscillator consists of an amplifier and a frequency selective network connected in a positive-feedback loop, as shown in *Figure 2-14* which is a block diagram of a positive-feedback loop. There is no input signal in a practical oscillator circuit. The input signal is shown in *Figure 2-14* to explain the principle of operation of a positive-feedback loop.

The gain of a positive-feedback circuit is A' = A/(1 - AB). Unlike the negative-feedback loop shown in *Figure 2-2*, the feedback signal of a positive-feedback loop is summed to the input signal with a positive sign. The loop gain of a positive feedback circuit is -AB.

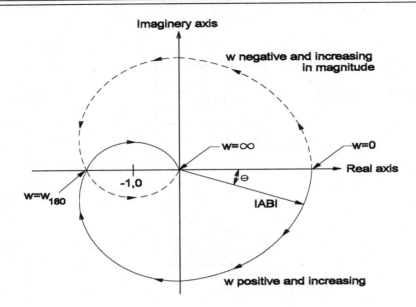

Figure 2-15. The Nyquist plot.

Positive feedback increases the gain of an amplifier, and sometimes called regenerative feedback. Positive or regenerative feedback increases the distortion content of the output signal of an amplifier. Positive feedback is rarely used in amplifier designs because regenerative feedback increases the distortion as well as the gain of an amplifier.

Positive feedback can be used in the design of frequency selective circuits. The bass and treble boost circuits of an audio amplifier are often positive feedback circuits.

Stability

When the feedback network is resistive, the feedback factor, B, is constant regardless of the frequency of the input signal. When the feedback network is capacitive or inductive, the feedback factor is dependent upon the frequency of the input signal. The closed-loop transfer function is:

$$A'(s) = A(s)/[1 + A(s)B(s)]$$

The loop gain, $A(s)B(s)$, varies with frequency and it determines the stability or instability of the feedback amplifier. When the loop gain is positive, the feedback is negative. When the loop gain is negative, the feedback is positive.

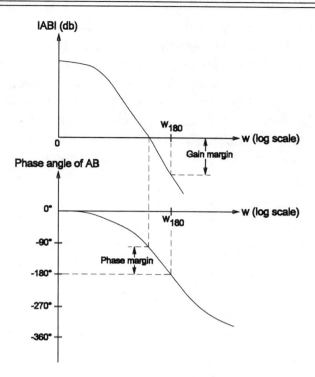

Figure 2-16. The Bode plot.

The Nyquist plot is a formal test for stability. It is a polar plot of loop gain with frequency used as a parameter, as shown in *Figure 2-15*. The radial distance is the magnitude of the loop gain, AB. The angle is the phase angle, theta. The solid-line plot is for positive frequencies. The dotted-line plot is for negative frequencies.

The Nyquist plot intersects the negative real axis at the frequency w_{180}. When this intersection occurs to the left of the point (-1,0), the magnitude of the loop gain at this frequency is greater than unity and the amplifier is unstable. When the intersection is to the right of the point (-1,0), the amplifier is stable. When the Nyquist plot encircles the point (-1,0), the amplifier is unstable.

The frequency response and stability of an amplifier are determined directly by its poles. When the poles of an amplifier are in the left half of the s-plane, the amplifier is stable. A pair of complex conjugate poles on the *jw* axis of the s-plane gives rise to sustained sinusoidal oscillations. The amplifier is therefore unstable. Poles in the right half of the s-plane give rise to growing oscillations. The amplifier is therefore unstable.

The Bode plot, as shown in *Figure 2-16*, is a simple method for determining whether a circuit is stable or unstable by examining the loop gain, AB, as a function of frequency. A feedback amplifier is stable if its loop gain magnitude at w_{180} is less than unity. The difference is known as the gain margin. A feedback amplifier is also stable if the magnitude of its loop gain is less than unity at a frequency less than 180 degrees phase shift, w_{180}. The phase margin is the difference between degrees and the phase angle at which the magnitude of the amplifier loop gain is unity. The amplifier is considered unstable if, at the frequency of unity loop gain magnitude, the phase lag is in excess of 180 degrees.

Problems

Problem 2-1. What are some effects of negative feedback on a circuit?

Problem 2-2. Prove that A' = A/(1 + AB) = 1/B if and only if AB >> 1.

Problem 2-3. What is the midband gain of a negative-feedback amplifier if B = 0.1 and the open-loop midband gain, Am, is 1000?

Problem 2-4. What is the upper 3 dB frequency of a negative-feedback circuit if the open-loop upper 3dB frequency is 10 kHz, Am = 1000 and B = 0.1?

Problem 2-5. What is the new lower 3dB frequency of a negative-feedback circuit if the open loop lower 3dB frequency is 100 Hz, Am = 1000 and B = 0.1?

Problem 2-6. Calculate the closed-loop gain, A', the closed-loop input resistance, R'$_{IN}$, and the closed-loop output resistance R'$_O$, of a series-shunt feedback circuit, if the open-loop gain, A, is 100, the open-loop input resistance R$_{IN}$, is 10 kohm and the open-loop output resistance, R$_O$, is 100 ohms. The feedback factor, B, is 0.1.

Problem 2-7. Repeat *Problem 2-6* for a shunt-series feedback circuit.

Problem 2-8. Repeat *Problem 2-6* for a series-series feedback circuit.

Problem 2-9. Repeat *Problem 2-6* for a shunt-shunt feedback circuit.

Problem 2-10. What type of amplifier should be used in a series-shunt feedback circuit?

Problem 2-11. What type of amplifier should be used in a shunt-series feedback circuit?

Problem 2-12. What type of amplifier should be used in a series-series feedback circuit?

Problem 2-13. What type of amplifier should be used in a shunt-shunt feedback circuit?

Problem 2-14. What is the positive feedback gain of a circuit whose open-loop gain is 10 and whose feedback factor is 0.1?

Problem 2-15. Repeat *Problem 2-14* for B = 0.0 and for B = 0.06.

Problem 2-16. What are some effects of positive feedback on a circuit?

Problem 2-17. Where do the poles of a stable circuit lie?

Problem 2-18. Define gain margin.

Problem 2-19. Define phase margin.

Problem 2-20. When is an amplifier considered stable according to a Nyquist plot?

Problem 2-21. When is an amplifier considered stable according to a Bode plot?

Chapter 3

Sinusoidal Oscillators

There are two approaches to designing a sinewave generator. In the first approach, the output of a nonlinear oscillator, which generates square and triangular waveforms, is fed to a sinewave shaper. A sinewave shaper usually consists of diodes and resistors. The nonlinear oscillator is also called a waveform generator or a function generator.

The second approach is to design a sinusoidal oscillator.

A sinusoidal oscillator is an amplifier with a positive-feedback loop that contains a frequency selective network, as shown in *Figure 3-1*. A sinusoidal or linear oscillator generates an output signal of a specific frequency which is determined by the values of the frequency selective components in the feedback network.

Despite the name, all oscillators are nonlinear circuits. Some form of nonlinearity must be employed to provide control of the amplitude of the output sinewave.

The positive feedback is accomplished by inductive, capacitive or resistive coupling between the output and the input of the amplifier. Oscillators may be designed to generate a wide range of frequencies by selecting components of the proper values.

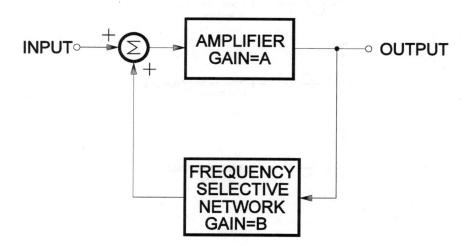

Figure 3-1. *Block diagram of a sinusoidal oscillator.*

Most oscillators may be classified as either tuned inductive-capacitive (LC) type or relaxation resistance-capacitance (RC) types. An electromechanical oscillator generates an output with a very stable frequency characteristic. A constant frequency reference, such as a crystal, is used as a constant source of vibration.

The feedback loop of *Figure 3-1* provides a sinusoidal oscillation at frequency w_o, if the phase of the loop gain at w_o is zero and the magnitude of the loop gain, AB, at w_o is unity. This is known as the Barkhausen criterion, and it may be written as $A(jw_o)B(jw_o) = 1$. This criteria must be satisfied at only one frequency if a sinusoidal waveform is required. If the criteria is satisfied at more than one frequency, a non-sinusoidal waveform is obtained.

Mode of Operation

A sinusoidal oscillator consists of an amplifier and a frequency-selective positive-feedback network, shown in *Figure 3-1*. The amplifier section is designed using active elements such as transistors or operational amplifiers. The feedback network is designed using passive elements such as resistors, capacitors and inductors. Inductors are the last choice because they tend to be bulky and expensive.

The feedback network feeds a portion of the amplifier output back to its input, in phase with its input. Signal regeneration results from positive feedback.

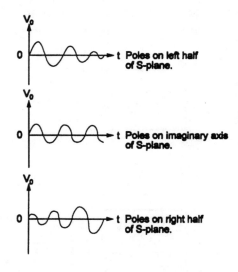

Figure 3-2. Pole locations and output waveforms.

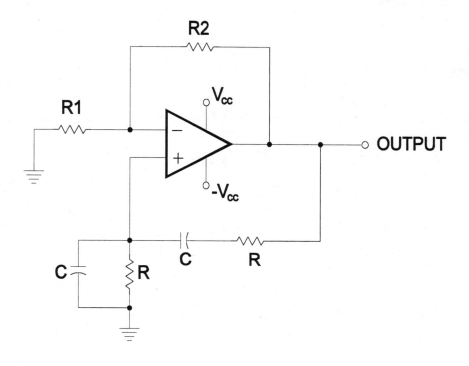

Figure 3-3. Basic operational-amplifier Wien-Bridge oscillator.

An oscillator can be regarded as a circuit whose response to an impulse input is a sinusoidal output. You may expect that an initial excitation is necessary to start the oscillation. This is not necessarily the case because there are spurious signals present in all practical oscillator circuits. Resistors generate thermal noise, and transistors generate shot noise. A noise signal can be regarded as a series of impulses. The required impulse input is therefore present, and the oscillator spontaneously oscillates as soon as power is applied to it.

A steady sinusoidal oscillation occurs if the poles of the oscillator circuit transfer function are placed on the imaginary axis of the s-plane. The pole positions depend on the component values of the feedback network, and on the open-loop gain of the amplifier. These parameters can vary with temperature and humidity. If the poles move to the left half of the s-plane, oscillations will cease because the circuit becomes stable. This can be avoided if the poles of the oscillator circuit transfer function lie on the right half of the s-plane. *Figure 3-2* shows the different output waveforms for different pole locations.

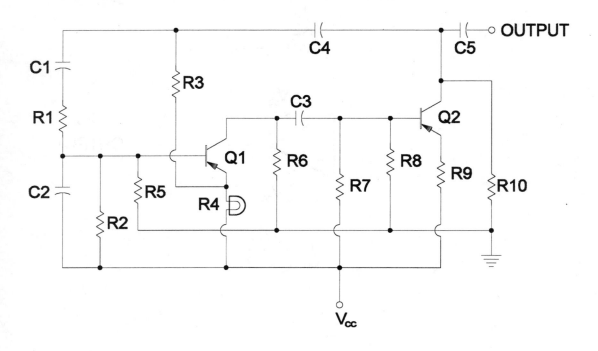

Figure 3-4. *Transistor Wien-Bridge oscillator.*

When the poles lie in the right half of the s-plane, the amplitude of oscillation tends to increase exponentially with time. In theory, the amplitude of the oscillation increases indefinitely. The practical amplifier has a finite signal handling capacity. Therefore, as the level of oscillation increases beyond a limit, the amplification decreases. A dynamic balance is achieved such that the level of oscillation is high enough to maintain a loop gain of unity. If the oscillation decreases, the loop gain increases, and the oscillation is increased. If the oscillation increases, the loop gain decreases, and the oscillation is decreased. The net result is a steady oscillation.

There are two important specifications for an oscillator circuit. These are the frequency of oscillation and the condition that will ensure sustained oscillation.

The frequency of oscillation can be determined by finding the imaginary part of the poles of the circuit transfer function. The condition that will ensure sustained oscillation can be deduced from the real part of the poles of the circuit transfer function.

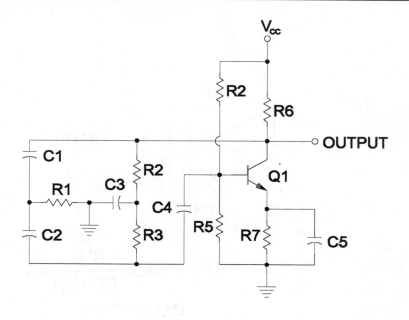

Figure 3-5. Twin-T bridge oscillator.

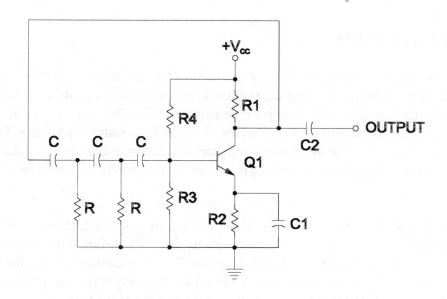

Figure 3-6. Transistor phase-shift oscillator.

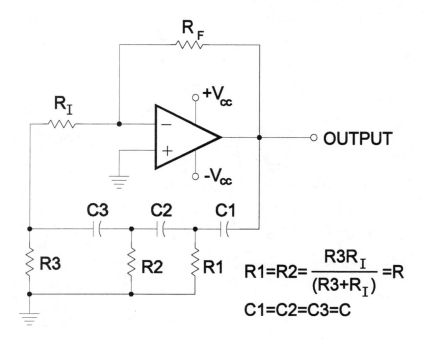

$$R1=R2= \frac{R3R_I}{(R3+R_I)} =R$$

$$C1=C2=C3=C$$

Figure 3-7. Operational-amplifier phase-shift oscillator.

Wien-Bridge Oscillator

A Wien-Bridge oscillator is shown in *Figure 3-3*. The operational amplifier is connected in the non-inverting amplifier configuration with a closed-loop gain of $(1 + R2/R1)$. An RC network is connected in the feedback path of the positive-gain amplifier. The frequency of oscillation, w_o, is $w_o = 1/RC$. The magnitude of the loop gain must be unity for sustained oscillation. This is achieved if $R2/R1 = 2$. To ensure that oscillations will start, $R2/R1$ should be slightly greater than 2. The amplitude of oscillation can be determined and stabilized by a nonlinear control network (not shown in *Figure 3-3*).

A transistor Wien-Bridge oscillator is shown in *Figure 3-4*. Each transistor stage produces an output that is 180 degrees out of phase with its input. Therefore, the two stages combined produce an output that is in phase with its input. This is necessary for positive feedback in the base circuit. Negative feedback is obtained by inserting a portion of the feedback into the emitter circuit of the input stage. Resistors R4 and R9 provide temperature stabilization. Positive feedback is controlled by the bridge reactive arms R1C1 and R2C2.

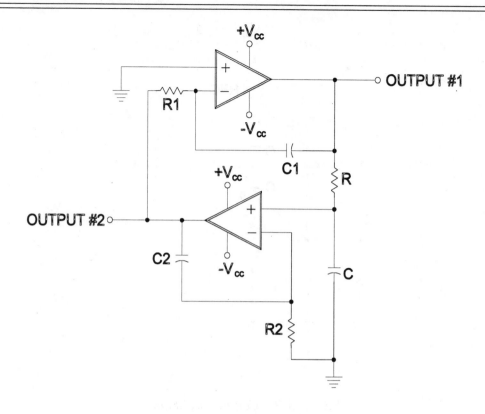

Figure 3-8. Quadrature oscillator.

A twin-T bridge oscillator is shown in *Figure 3-5*. The DC blocking RC network consists of resistors R1, R2 and R3 and of capacitors C1, C2 and C3. The transistor is biased by resistors R4, R5, R6 and R7. Capacitor C4 couples the oscillator signal to the base of the transistor and it also blocks DC. Capacitor C5 bypasses AC signals and prevents degeneration.

Phase-Shift Oscillator

Sustained oscillation is only possible if the feedback is positive and the amount of feedback is large enough for a loop gain of unity. In a phase-shift oscillator, the amplifier provides an output that is 180 degrees out of phase with its input. The output is fed back to the input through a phase-shift network that produces an additional 180 degree phase shift at the frequency of oscillation. A transistor phase-shift oscillator is shown in *Figure 3-6* and an operational-amplifier phase-shift oscillator is shown in *Figure 3-7*. The operational-amplifier is configured as an inverting voltage amplifier.

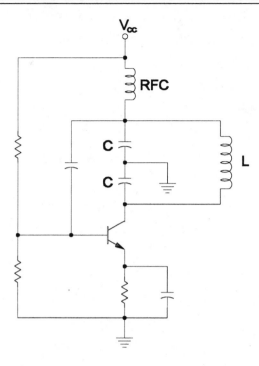

Figure 3-9. *Colpitts oscillator.*

The gain of the amplifier must be at least 29 for the circuit to oscillate. The frequency of oscillation is $w_o = 1/SQRT(6)RC$; that is, $f_o = 1/2 * PI * SQRT(6)RC$, because $f_o = w_o/2 * PI$.

The feedback network is a three-section or third-order RC ladder network. Three is the minimum number of sections; that is, the lowest order that is capable of producing 180 degrees phase shift at a finite frequency.

If the gain of the amplifier is less than 29, the circuit will not oscillate. If the gain is over approximately 32, the output is a clipped sinewave. The output is an undistorted sinewave if the gain of the amplifier stage is $29 < A_v < 32$.

Quadrature Oscillator

The quadrature oscillator provides two sinusoidal outputs with a 90 degree phase difference. An operational-amplifier quadrature oscillator is shown in *Figure 3-8*.

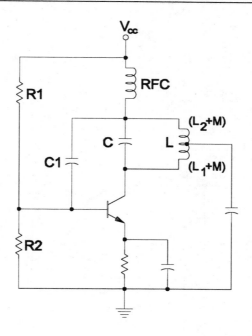

Figure 3-10.   Hartley oscillator.

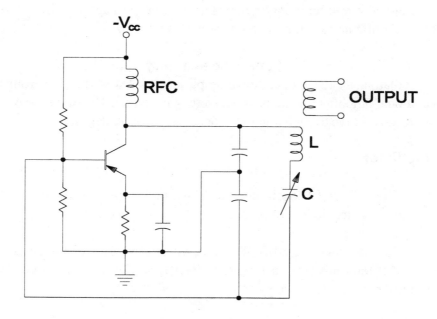

Figure 3-11.   Clapp oscillator.

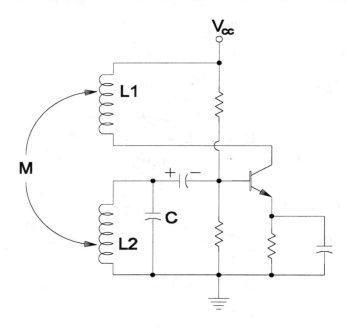

Figure 3-12. Basic Armstrong oscillator.

The upper operational-amplifier is configured as an inverting voltage amplifier. The lower operational-amplifier is configured as a non-inverting voltage amplifier.

If R1 = R2 = R, and if C1 = C2 = C, then the frequency of oscillation is w_o = 1/RC. At the frequency of oscillation, the lower operational-amplifier operates as a non-inverting integrator and the upper operational-amplifier functions as an inverting integrator. Therefore, the two output waveforms are displaced relative to each other by a phase angle of 90 degrees.

Colpitts Oscillator

The Colpitts oscillator is an LC oscillator, as shown in *Figure 3-9*. The tuned LC circuit is a parallel resonant tank circuit. There is no tap on the oscillator coil of a Colpitts oscillator.

The frequency of oscillation is w_o = 1/SQRT(LC), where C = C1C2/(C1 + C2). Sustained oscillation is obtained if the transistor current gain, B, is B = C2/C1. If C1 = C2, then sustained oscillation occurs if beta is at least unity.

The Colpitts oscillator operates well into a class-C condition for transistors that have normal values of beta. Capacitors C1 and C2 are usually of equal capacitance.

Hartley Oscillator

The Hartley oscillator is also an LC oscillator, as shown in *Figure 3-10*. The Hartley oscillator is used to produce a sinewave output of constant amplitude and fairly constant frequency in the medium and high radio frequency ranges. The Hartley oscillator is generally
used as a local oscillator in receivers, as a signal source in signal generators, and as a variable-frequency oscillator for general use.

The Hartley oscillator uses a tapped coil in the oscillator tuned circuit. The frequency of
oscillation is $w_o = 1/SQRT(LC)$, where $L = L1 + L2 + 2M$. The frequency of oscillation may also
be written as $f_o = 1/2 * PI * SQRT(LC)$. For sustained oscillation, the transistor beta must be $B = (L2 + M)/(L1 + M)$.

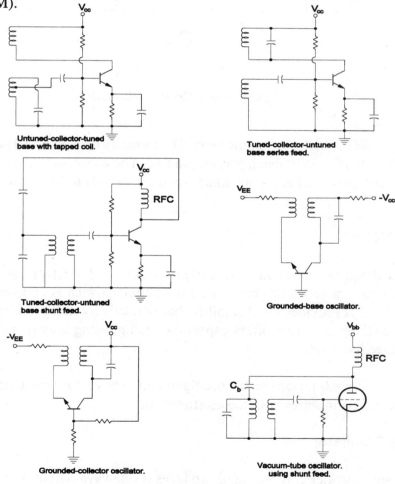

Figure 3-13. Versions of the Armstrong oscillator.

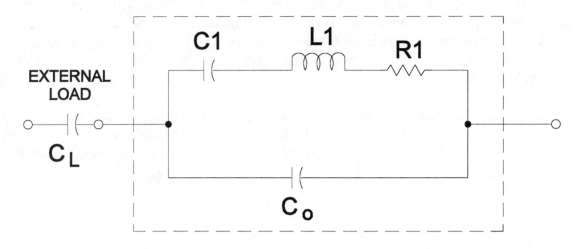

Figure 3-14. Crystal schematic.

Usually, the tap on the coil is not at the center. It is usually located on 10 percent of the coil. The transistor beta is usually much greater than 10. The automatic action of a class-C bias clamp lowers the effective beta of the transistor until a stable point of sinusoidal oscillation is reached.

Clapp Oscillator

The Clapp oscillator is a variation of the Colpitts oscillator. The Clapp oscillator is shown in *Figure 3-11*. It uses a series-resonant tuned tank circuit which is coupled to the positive-feedback loop. This provides good stability that is relatively independent of transistor parameters. The Clapp oscillator offers capacitive tuning, using only one capacitor, without affecting the feedback ratio.

The Clapp oscillator has no taps on the oscillator coil. It has a third capacitor in the LC tuned tank circuit which is in series with the oscillator coil.

Armstrong Oscillator

The Armstrong or tickler coil oscillator produces a sinewave output of constant amplitude and fairly constant frequency within the RF range. It is usually used as a local oscillator in

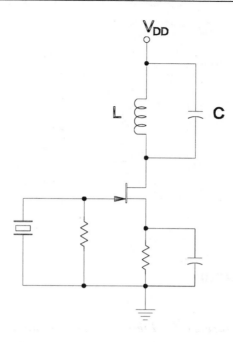

Figure 3-15. Miller crystal oscillator.

receivers, as a signal source in signal generators, and as a variable-frequency oscillator over the medium and high frequency ranges. A basic Armstrong oscillator is shown in *Figure 3-12*.

The tank circuit is in either the base or the collector circuits. The tuned-base and the tuned-collector circuits are two versions of the Armstrong oscillator, shown in *Figure 3-13*.

The Armstrong oscillator uses an air-core transformer in parallel with a tuning capacitor. The positive feedback is determined by the relative polarity of the coils. The amount of feedback is controlled by the mutual inductance, M, between L1 and L2.

As soon as the Armstrong oscillator is powered up, the collector current increases, causing an increased flux in the collector coil. A secondary voltage in the base winding is caused by the increased flux. The base voltage polarity drives the transistor positive and charges the tuning capacitor C. When the capacitor voltage becomes negative, the tank circuit starts to oscillate. The negative base voltage reduces the collector current. A voltage is induced in the base winding, making the current reduction a cumulative process. After a few cycles, the amplitude of the oscillation reaches its final value.

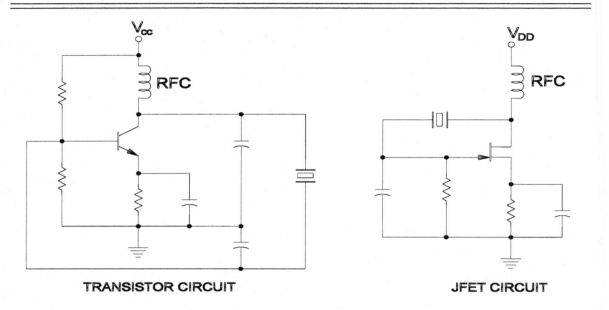

TRANSISTOR CIRCUIT   JFET CIRCUIT

Figure 3-16. Pierce crystal oscillator.

A class-C bias is necessary to maintain a sinusoidal waveform and to prevent excessive currents from destroying the transistor. The oscillator stabilizes at a point where the circuit losses equal the AC power output of the transistor.

The emitter resistor limits the initial current surge to a safe value when the circuit is first powered up. The Armstrong oscillator uses an oscillator coil with two separate windings containing four terminals.

When the tuning capacitor is in the base circuit, the collector winding is called the tickler, or feedback winding. When the tuning capacitor is in the collector circuit, the base winding becomes the feedback winding. In the latter configuration, the tuning capacitor is connected to the power supply.

In a series-feed circuit, DC collector current flows through at least part of the oscillator coil. In a shunt or parallel feed, DC collector current is kept out of the coil by blocking capacitor C_b. An RF choke must be placed in series with the collector to keep the collector at a high AC impedance with respect to ground.

To prevent a transistor from loading the oscillator tank circuit, a tap on the tank coil for the transistor connection is used.

Crystal Oscillator

Crystalline substances such as quartz change shape when an EMF is impressed on the crystal. Conversely, when the crystal is subject to mechanical stress, an EMF is produced across the surface of the crystal. The interrelation of mechanical and electrical stress in crystals is known as the piezoelectric effect.

A crystal can be represented schematically, as shown in *Figure 3-14*. The series components R1, C1 and L1 are called the motional elements. The motional elements L1 and C1 determine the series resonant frequency of the crystal. Resistor R1 determines the quality factor, Q, of the resonator. The series resonant frequency is $w_o = 1/SQRT(L1C1)$. Capacitor C1 is very small, on the order of one picofarad or less. Inductor L1 is large, on the order of several millihenries. Capacitor C1 may be specified when the crystal is pulled. Inductor L1 cannot be specified directly; it is indirectly specified by the frequency of operation.

A shunt capacitance, C_o, appears across the series motional elements. It is a function of the crystal holder, called the parallel capacitance.

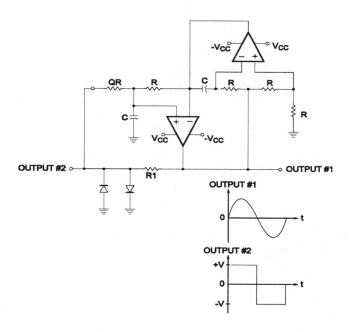

Figure 3-17. Active-filter tuned oscillator.

The load capacitance, C_L, takes the form of an intentional external varactor, used for frequency pulling or the stray capacitance of an oscillator circuit.

Crystals can oscillate in two modes, fundamental and overtone. In the fundamental mode, the crystal oscillates at its natural resonant frequency. The frequency of oscillation in the fundamental mode depends upon the crystal's dimensions, the way it has been cut, and temperature.

In the overtone mode, the crystal oscillates at a frequency that is approximately an integer multiple of its fundamental frequency. The overtone frequency is not a harmonic of the fundamental frequency.

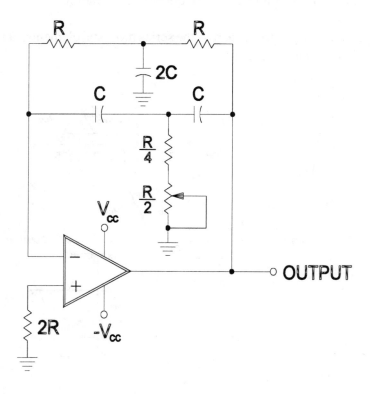

Figure 3-18.   Twin-T sinewave oscillator.

Rochelle salt crystals are used in microphones and phonograph pickups. Quartz crystals are used in radio-frequency (RF) oscillators. When a quartz crystal is properly cut and ground, it acts like a parallel-resonant circuit of high Q and can be used as a tank circuit in oscillators. The resonant frequency of a crystal is inversely proportional to its size and thickness. The orientation of the crystal cutting determines the electrical properties of the crystal. Some cuts have a great deal of waste and are therefore more expensive.

A Miller crystal oscillator is shown in *Figure 3-15*. The feedback occurs solely through the inter-electrode capacitance of the JFET.

The Pierce crystal oscillator shown in *Figure 3-16* is a version of the Colpitts oscillator. The crystal operates in the parallel mode in the Pierce oscillator. The Pierce oscillator is very convenient for transmitters that operate crystal-controlled on several channels. The different crystals can be switched in without having to retune the oscillator circuit.

Crystals may be used in oscillators with series resonant circuits, and in oscillators with parallel resonant circuits. The positive-feedback loop is a series-feedback path and a parallel-feedback path, respectively.

A variable capacitor may be placed in parallel with the crystal to fine-tune the crystal's frequency of oscillation. The adjustment possible is about 100 Hertz per 1,000,000 Hertz.

Operational-Amplifier Oscillator

An active-filter tuned oscillator is shown in *Figure 3-17*. A high-Q bandpass filter is connected in a positive-feedback loop with a limiter. The output of the bandpass filter is a sinewave whose frequency is equal to the center frequency of the bandpass filter. The sinewave is fed to the limiter which then outputs a squarewave. The squarewave levels are determined by the limiting levels of the limiter. The squarewave is fed back to the bandpass filter which filters out the harmonics to generate a sinusoidal output at the fundamental frequency. The purity of the output sinewave is a function of the selectivity or Q factor of the bandpass filter. The frequency, amplitude, and distortion of the output sinewave are independently controlled.

A twin-T operational-amplifier sinewave oscillator is shown in *Figure 3-18*. A twin-T network is used in the positive-feedback loop of the oscillator. When the twin-T network is slightly detuned, the phase of the feedback is shifted 180 degrees, and the circuit oscillates at $f_o = 1/2 * PI * RC$. The potentiometer detunes the twin-T network. The sinewave becomes distorted when the potentiometer resistance is too low. There is no output when the potentiometer resistance is too high.

Problems

Problem 3-1. Name two types of oscillators.

Problem 3-2. What is an electromechanical oscillator? Give an example.

Problem 3-3. What is the Barkhausen criterion for oscillation?

Problem 3-4. What initial excitation is required for an oscillator to oscillate?

Problem 3-5. Name some spurious signals.

Problem 3-6. What is a noise signal?

Problem 3-7. Where are the poles of a transfer function of an oscillator?

Problem 3-8. What determines the pole positions of an oscillator?

Problem 3-9. What are two important specifications of an oscillator?

Problem 3-10. How are the two specifications determined?

Problem 3-11. Calculate the frequency of oscillation of a Wien-bridge oscillator if R = 100 kohm and C = 0.001 uF.

Problem 3-12. What is the condition for sustained oscillation of a phase-shift oscillator?

Problem 3-13. Calculate the frequency of oscillation of a phase-shift oscillator if R = 100 kohm and C = 0.001 uF.

Problem 3-14. Calculate the frequency of oscillation of a quadrature oscillator if R = 100 kohm and C = 0.001 uF.

Problem 3-15. Compare the tank circuits of the Colpitts, Clapp, Hartley and Armstrong oscillators.

Problem 3-16. Calculate the frequency of operation of a Colpitts oscillator if C = 0.001 uF and L = 10 mH.

Problem 3-17. Calculate the frequency of oscillation of a Hartley oscillator if C = 0.001 uF and L = 10 mH.

Problem 3-18. Where are Armstrong oscillators used?

Problem 3-19. What is the piezoelectric effect?

Problem 3-20. What determines the series resonant frequency of a crystal?

Problem 3-21. What determines the Q factor of a resonator?

Problem 3-22. What is the crystal frequency of oscillation if C1 = 1 pF and L1 = 10 mH?

Problem 3-23. How is the series capacitance of a crystal specified?

Problem 3-24. What is the shunt capacitance?

Problem 3-25. What are the two modes of crystal operation?

Problem 3-26. Discuss the fundamental mode.

Problem 3-27. Discuss the overtone mode.

Problem 3-28. Where are Rochelle salts used?

Problem 3-29. Where are quartz crystals used?

Problem 3-30. Why is the Pierce oscillator used in transmitters that operate on several crystal-controlled channels?

Problem 3-31. How are frequency, amplitude and output distortion controlled in an active-filter tuned oscillator?

Problem 3-32. What is the frequency of oscillation of an operational-amplifier twin-T oscillator if R = 100 kohm and C = 0.001 uF?

Problem 3-33. What happens if the potentiometer resistance of an operational-amplifier twin-T oscillator is too low?

Problem 3-34. What happens if the potentiometer resistance of an operational-amplifier twin-T oscillator is too high?

Chapter 4

Designing Sinusoidal Oscillators

If an oscillator that is operating at high power fails, oscillation stops. The circuit draws excessive current. The amplifier must be protected from excess current drain by a fuse, a resistor or a panel light in the DC supply circuit. The panel lamp resistance increases when the lamp current increases.

Power supply fluctuations can produce variations in the frequency and output level of an oscillator. A regulator is often used to provide a constant DC supply for the oscillator circuit.

Temperature changes can cause mechanical contractions and expansions within the coil and capacitor, causing frequency and output variations. Many oscillators have temperature-compensation capacitors in parallel with the tank circuit. These capacitors are available with different temperature coefficients.

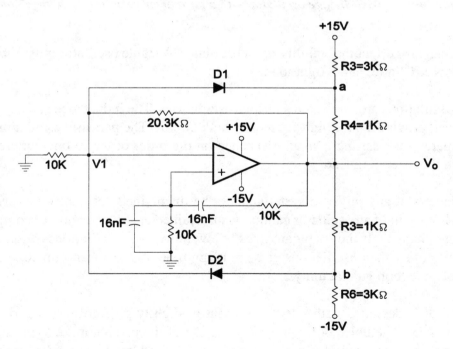

Figure 4-1. A Wien-Bridge oscillator with a limiter used for amplitude control.

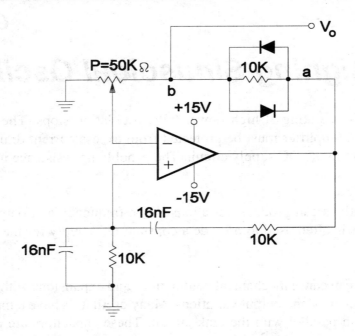

Figure 4-2. A Wien-Bridge oscillator with a parameter variation amplitude control.

Humidity can also affect the stability of an oscillator. A stable oscillator must have a high Q factor, a low L/C ratio, and a light load.

Parasitic oscillations are unwanted spurious oscillations. The inductance and capacitance of the leads can provide a very high-frequency tank circuit. The parasitic oscillations are usually eliminated by inserting a small resistance on the order of ten to one hundred ohms in series with the offending lead.

The parameters of any physical system cannot be maintained constantly for any length of time. The loop gain of an oscillator cannot be maintained at unity because of component and temperature changes. If the loop gain goes below unity then oscillations cease. If the loop gain exceeds unity then oscillations grow in magnitude. A nonlinear gain control circuit is used to keep the loop gain at unity.

The oscillator is designed with a loop gain that is slightly greater than one to ensure that oscillations start. The poles of the transfer function of the oscillator are in the right half of the s-plane. When the oscillator output reaches the desired level, the nonlinear gain control circuit causes the loop gain to be reduced to and maintained at unity. The poles are pulled

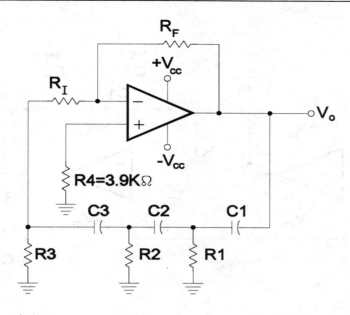

Figure 4-3. Phase-shift oscillator.

back to the imaginary or *jw* axis of the s-plane. The resulting output is sustained oscillations at the desired amplitude. If the loop gain goes below one, the output amplitude decreases. The nonlinear gain control circuit detects this and increases the gain loop to exactly unity.

A limiter circuit is often used as a nonlinear gain control. *Figure 4-1* is a schematic of a Wien-Bridge oscillator employing a limiter gain control circuit. The output increases until it reaches the preset limiter level. The limiter keeps the output amplitude constant. The limiter should be "soft" to minimize nonlinear distortion of the output.

The amplitude may also be stabilized by detecting the output, and converting the output level to a DC voltage. The DC voltage is compared to a preset value. The comparator output is used to adjust the resistance that determines the loop gain. This technique of gain control is shown in *Figure 4-2*.

The circuit of *Figure 4-1* uses a symmetrical feedback limiter consisting of diodes D1 and D2 and resistors R3, R4, R5 and R6. At the positive peak of the output voltage, the voltage at node *b* exceeds V1 and diode D2 conducts. The positive peak of the output voltage is clamped to a value determined by R5, R6 and the negative power supply. Similarly, the

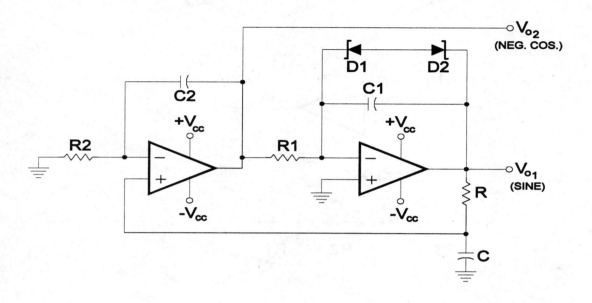

Figure 4-4. Quadrature oscillator.

negative peak of the output of the output voltage is clamped to the value that causes diode D1 to conduct. For a symmetrical output waveform, R3 = R6 and R4 = R5.

The circuit of *Figure 4-2* uses a parameter variation mechanism as a gain control. Potentiometer *P* is adjusted until oscillations just start to grow. The diodes start to conduct when the oscillations grow. The effective resistance between *a* and *b* decrease. Equilibrium is reached at the output amplitude that causes the loop gain to be exactly unity. The amplitude of the output voltage can be varied by adjusting potentiometer *P*.

Designing a Phase-Shift Oscillator

The phase-shift oscillator shown in *Figure 4-3* is designed to oscillate at 1000 Hz. For oscillation to occur, the open-loop gain of the oscillator, A, must exceed 30. The open-loop gain of the circuit is $A = R_F/R_I$. Let $R1 = R2 = R_I R3/(R_I + R3) = R$. The frequency of oscillation is $f_o = 1/2 * PI * SQRT(6)RC$. If C1 = C2 = C3 = 0.047 uF, then 1000 = 1/6.28 * R * 0.047 uF, from which R = R1 = R2 = 1,382 ohms. Also, let R_I = 8,200 ohms. 1,382 = 8,200R3/ (8,200 + R3), from which R3 = 1,406 ohms. Since $A = R_F/R_I$, R = 246 kohms. Finally, let R4 = 3,900 ohms.

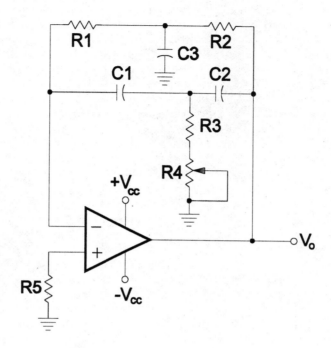

Figure 4-5. Twin-T oscillator.

Designing a Quadrature Oscillator

The quadrature oscillator shown in *Figure 4-4* is designed to oscillate at 1000 Hz. $f_o = 1/2$ * PI * RC. Let C = C1 = C2 = 0.047 uF. 1000 =1/6.28 * R * 0.047 uF, from which R = R1 = 3,386 ohms. For oscillations to start, R2 < R; therefore let R2 = 3,300 ohms.

Zener diodes D1 and D2 form a limiter for circuit gain stability. The oscillations are therefore stabilized and the output amplitude is constant.

Designing a Twin-T Oscillator

The twin-T oscillator of *Figure 4-5* is designed to oscillate at 1000 Hz. $f_o = 1/2$ * PI * RC. Let C = C1 = C2 = 0.047 uF. Substituting into the equation for f_o, R = R1 = R2 = 3,388 ohms. Therefore, R3 = 847 ohms, R4 = 1694 ohms, and R5 = 6,776 ohms. Finally, C3 = 0.094 uF.

Problems

Problem 4-1. How can the amplifier of an oscillator be protected from excess current drain?

Problem 4-2. What causes fluctuations in the frequency and amplitude of the output of an oscillator?

Problem 4-3. How can fluctuations in the output of an oscillator be prevented?

Problem 4-4. What three factors determine the stability of an oscillator?

Problem 4-5. What are parasitic oscillations?

Problem 4-6. What causes parasitic oscillations?

Problem 4-7. How can parasitic oscillations be eliminated?

Problem 4-8. What is the purpose of a nonlinear gain control circuit in an oscillator?

Problem 4-9. What are two types of nonlinear gain control circuits?

Problem 4-10. Design a phase-shift oscillator that oscillates at 500 Hz.

Problem 4-11. Design a quadrature oscillator that oscillates at 500 Hz.

Problem 4-12. Design a twin-T oscillator that oscillates at 500 Hz.

Chapter 5

Schmitt Trigger

In a sinusoidal oscillator, the period of the cycle is usually determined by the resonant frequency of a tuned circuit. The conducting and nonconducting times of the amplifier are determined by the class-C bias conditions. The charge and discharge times of a capacitor determine the period of the cycle in a non-sinusoidal oscillator.

The capacitor in *Figure 5-1* begins to charge when power is applied at time zero. The capacitor voltage increases as a function of the time constant $t = RC$, until the capacitor voltage reaches an asymptotic value, E, as shown in *Figure 5-2*.

The Schmitt trigger is an emitter coupled circuit, as shown in *Figure 5-3*. It is a bistable circuit which can remain indefinitely in either of its two stable states; that is, Q1 off and Q2 on, or Q1 on and Q2 off. The state that a Schmitt trigger is in is selected by a voltage level, not by a trigger pulse; therefore, the Schmitt trigger is used as a "squarer," or as a voltage level detector.

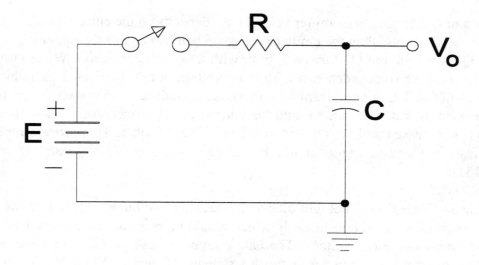

Figure 5-1. *RC charging circuit.*

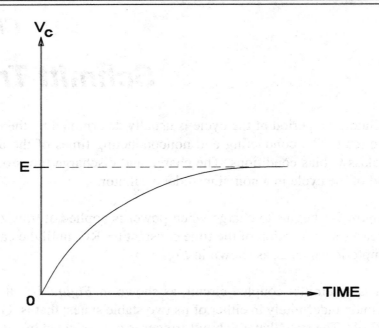

Figure 5-2. RC circuit response.

When Q2 is on or conducting, the emitter voltage, V_E, depends on the current in Q2. Q1 is in cutoff because Vin is less than the emitter voltage. When Vin equals the upper trip level voltage, V1, Q1 turns on and Q2 turns off, both with a fast trigger action. When current flows in Q1, its collector voltage decreases. The base voltage of Q2 also decreases, reducing the current flow in Q2. The common emitter voltage is also reduced, the voltage on the base of Q1 is increased as Vin is increased, and the current in Q1 is increased. The collector voltage of Q1 is increased and the current in Q2 is reduced further. This process repeats itself with a fast, regenerative trigger action. For all input voltages, V1 or greater, Q1 conducts and Q2 is cutoff.

If Vin is returned to V1, Q1 does not turn off and Q2 does not turn on. For Q1 to turn off, the current in Q1 must be reduced considerably which would increase the base voltage on Q2 and lower the common emitter voltage. The base-emitter voltage on Q2 equals the cutin voltage of Q2. This happens when Vin is much less than V1, that is, Vin = V2. V2 is the lower trip level voltage. The Schmitt trigger has hysteresis, as shown in the transfer curve in *Figure 5-3*.

The hysteresis voltage, V_H, is the difference between the upper and lower trip levels: $V_H = V1 - V2$. Without hysteresis, a noisy signal near a single trip point may cause the Schmitt trigger to switch randomly between its two states. For all input voltages, V2 or less, Q1 is cut off and Q2 conducts.

The Schmitt trigger is analyzed as follows:

1. Q1 off and Q2 on, calculate V_{B2}, $V1(UTL) = V_{B2} - 0.1$, V_E, I_E and dV_{C2}. Check that Q2 is in its active region.

2. Q1 on and Q2 off, $I_{C1} = I_E$ in terms of I_D at the Q2 turn-on point. I_D is the current in the divider R1 and R2. Write the divider chain equation in I_D and I_{C1}, substitute for I_{C1}. Solve for I_D. Calculate V_{B2} and $V2(LTL) = V_{B2} + 0.1$.

3. Calculate $V_H = V1 - V2$.

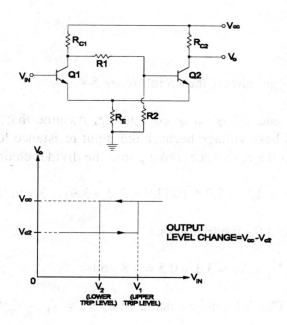

Figure 5-3. Schmitt trigger circuit and its transfer curve.

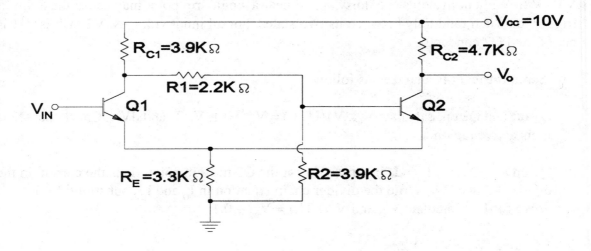

Figure 5-4. Schmitt trigger circuit for Example 5-1.

Example 5-1

Analyze the Schmitt trigger circuit shown in *Figure 5-4*.

Assume that Q1 is off and Q2 is on or conducting. Assume that the base current has a negligible effect on the base voltage because the input resistance looking into the base of Q2, is much higher than the resistance looking into the divider chain coupling Q1 to Q2.

$V_{B2} = R2Vcc/(R_{C1} + R1 + R2) = 3.9 * 10/(3.9 + 2.2 + 3.9) = 3.9$ volts

$V_E = V_{B2} - V_{BE}(\text{active}) = 3.9 - 0.6 = 3.3$ volts

For Q1 to be off, $V_{B1} < V_E + Va = 3.3 + 0.5 = 3.8$ volts.

The upper trip level (UTL), V1 can be calculated as $V1 = V_{B2} - V_{BE}(\text{active}) + Va$.

For a silicon transistor, $V1 = V_{B2} - 0.6 + 0.5 = V_{B2} - 0.1$ volts.

For a germanium transistor, $V1 = V_{B2} - 0.2 + 0.1 = V_{B2} - 0.1$ volts.

For any transistor, $V1 = V_{B2} - 0.1 = 3.9 - 0.1 = 3.8$ volts.

The upper trip level is a stable value which is independent of h_{FE}, temperature, or transistor type. This makes the Schmitt trigger useful as a voltage level detector.

$I_E = V_E/R_E = 3.3/3.3k = 1$ mA

$V_{RC2} = dV_{C2}$, because $V_{RC2} = 0$ when Q2 is cut off.

$V_{RC2} = I_{C2}R_{C2} = I_ER_{C2} = 1mA * 4.7k = 4.7$ volts

$V_{C2} = Vcc - V_{RC2} = 10 - 4.7 = 5.3$ volts

To verify that Q2 is not saturated, $V_{CE2} = V_{C2} - V_E = 5.3 - 3.3 = 2$ volts.

When Q1 conducts, Q2 must turn off, because the emitter voltage is fairly constant at 3.2 volts, while the collector voltage of Q1 has dropped by $I_{C1}R_{C1}$. The base voltage of Q2 also drops below the common emitter voltage.

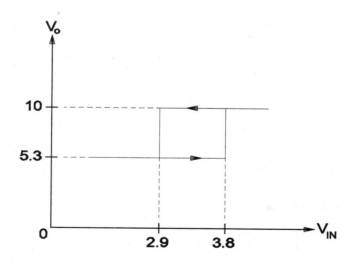

Figure 5-5. Transfer curve for Example 5-1.

The input voltage at which Q2 turns on again is calculated by reducing the current in Q1 until the collector voltage of Q1 has increased enough so that the base voltage of Q2 exceeds the emitter voltage by the cutin voltage, Va.

Q2 conducts when $V_E = V_{B2}$ - Va.

$I_E R_E = I_D R2$ - Va, where I_D is the current in divider R1, R2.

Since $I_E = I_{C1}$, $I_{C1} R_E = I_D R2$ - Va.

$3.3 I_{C1} = 3.9 I_D - 0.5$

$I_{C1} = (3.9 I_D - 0.5)/3.3$

Considering the divider chain R_{C1}, R1 and R2, $(I_{C1} + I_D)R_{C1} + I_D(R1 + R2) = Vcc$.

$3.9(I_{C1} + I_D) + 6.1 I_D = 10$

Substituting for I_{C1}, $3.9[\{(3.9 I_D - 0.5)/3.3\} + I] + 6.1I = 10$.

$1.18[3.9 I_D - 0.5] + 3.9 I_D = 6.1 I_D = 10$

$4.6 I_D - 0.59 + 10 I_D = 10$

$14.6 I_D = 10.59$

$I_D = 0.72$ mA

$V_{B2} = I_D R2 = 0.72 * 3.9 = 2.8$ volts

$V_E = V_{B2}$ - Va = 2.8 - 0.5 = 2.3 volts

$V2 = V_E + V_{BE}$(active) = 2.3 + 0.6 = 2.9 volts

It should be noted that $V2 = V_{B2} + 0.1 = 2.8 + 0.1 = 2.9$ volts.

The hysteresis voltage, $V_H = V1 - V2 = 3.8 - 2.9 = 0.9$ volts.

The transfer curve of this Schmitt trigger is shown in *Figure 5-5*. The Schmitt trigger can be used as a squarer, as shown in *Figure 5-6*. The symmetry of the output may be controlled with the DC level of the input waveform. Varying the DC level shifts the waveform up and down with reference to the fixed trip levels. This controls the points on the input waveform at which the output waveform changes state.

In an oscilloscope sweep trigger circuit, the trigger level control varies the DC level of the triggering waveform that is applied to the Schmitt trigger. The sweep trigger pulse is obtained by differentiating the output of the Schmitt trigger.

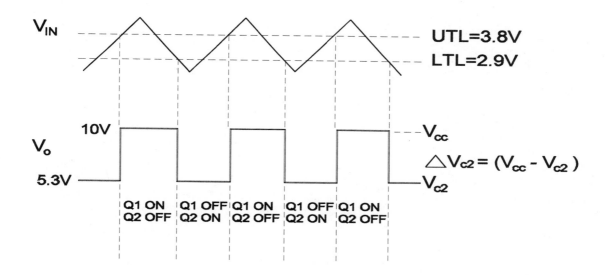

Figure 5-6. Schmitt trigger as a square waveform.

Problems

Problem 5-1. What determines the period of the cycle of a sinusoidal oscillator?

Problem 5-2. What determines the period of the cycle of a non-sinusoidal oscillator?

Problem 5-3. In the Schmitt trigger circuit shown in *Figure P5-1*, if Q2 is conducting and Q1 is turned on by a positive voltage of sufficient amplitude, what happens in the circuit?

Problem 5-4. Assume that Q2 is conducting, but not in saturation, and that it draws a very small base current. What is the base voltage of Q2?

Problem 5-5. What is the common emitter voltage if Q1 is off and Q2 is on?

Problem 5-6. What is the current in Q2? What is the output voltage in this state?

Problem 5-7. Is Q2 in saturation? Check V_{CE2}.

Problem 5-8. What voltage is required on the base of Q1 for Q1 to conduct? What happens to the circuit at this point?

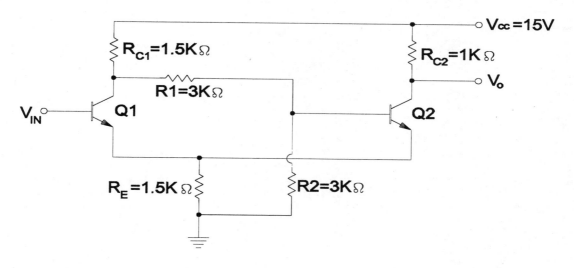

Figure P5-1. Schmitt trigger circuit for Chapter 5 problem set.

Problem 5-9. Is the turn-on voltage for Q1 affected by temperature or transistor type? Explain.

Problem 5-10. When current is flowing in Q1, how will the base voltage of Q2 compare with its original value? Show that Q2 must turn off when Q1 turns on. If the input voltage to Q1 is decreased, what happens to the base-emitter voltage of Q2?

Problem 5-11. Calculate the base voltage of Q2 at which Q2 turns on again.

Problem 5-12. What is the base voltage of Q1 and V2 or LTL corresponding to *Figure P5-1?*

Chapter 6

Multivibrators Using Transistors

Multivibrators are non-sinusoidal oscillators. They are used to store binary information, generate squarewaves and pulses, generate time delays, counting, and frequency division. A multivibrator is a two-stage amplifier, with its output connected to its input.

There are three types of multivibrators. The bistable multivibrator consists of two inverter transistor switches connected back-to-back. It has two stable states. The astable multivibrator consists of two off-gated transistor switches connected back-to-back. It has no stable state. The astable multivibrator is useful for generating pulses because it is a free-running squarewave oscillator. The monostable multivibrator consists of an inverter transistor switch connected back-to-back with an off-gated transistor switch. It has one stable state and requires a trigger input.

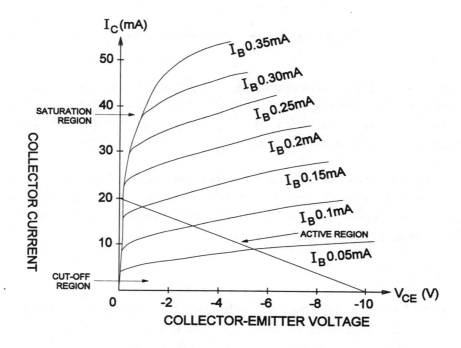

Figure 6-1. *Transistor operating regions.*

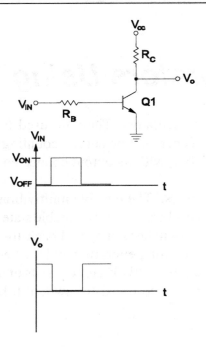

Figure 6-2. Basic transistor switch.

The transistor has three regions of operation, as shown in *Figure 6-1*. *Figure 6-1* is a typical transistor collector characteristics graph with a load line drawn. When a transistor is operating in the middle of its active region, it is operating in its linear mode. The transistor is usually biased to function as an amplifier or as an oscillator in its linear mode.

The transistor can function like a switch if it is biased at the extremes of the load line, as shown in *Figure 6-1*. When the transistor is not conducting, it is in cut-off mode and functions like an "open" switch. When the transistor is conducting heavily, it is in saturation mode and functions like a "closed" switch.

Saturation

Saturation is the ON state of the transistor. The basic transistor switch is shown in *Figure 6-2*. A transistor is saturated when the collector current has risen with increasing base drive, such that most of Vcc is dropped across Rc. Negligible voltage is dropped across the transistor; the transistor is therefore operating like a closed switch. In the saturated transistor, Vce < 0.4V (silicon), and Vce < 0.1V (germanium).

Ic(sat) = (Vcc - Vce)/Rc, and if Vce is very small, Ic(sat) = Vcc/Rc.

V_{BE}(sat) = 0.7V(silicon) and V_{BE}(sat) = 0.3V(germanium).

The internal resistance of a saturated transistor is 10 ohms or less. The power dissipated by the transistor is very low. The voltage levels of a saturated transistor switch are shown in *Figure 6-3*. The collector-base junction becomes forward-biased in a saturated transistor. It is reverse-biased when the transistor operates in its active region.

The DC current gain of a transistor is $h_{FE} = I_C/I_B$.

$I_B < I_B$(sat) and I_B(sat) = I_C(sat)/h_{FE}, and I_B = (Vin - V_{BE}(sat)/R_B, and $I_B > I_B$(sat).

If the base resistor is too large, the transistor will not saturate. If the DC current gain of a transistor is higher than its guaranteed minimum, it will go more deeply into saturation.

Cut-Off

The collector-base junction of a transistor is reverse-biased when it is operating in its cut-off region. The collector current can never be zero because there is leakage current flowing in the collector circuit. The threshold of cut-off is therefore defined as the point at which the emitter current becomes zero. For germanium transistors, a reverse-bias of 0.1V is required on the base-emitter junction. Silicon transistors require zero volts of reverse-bias on the

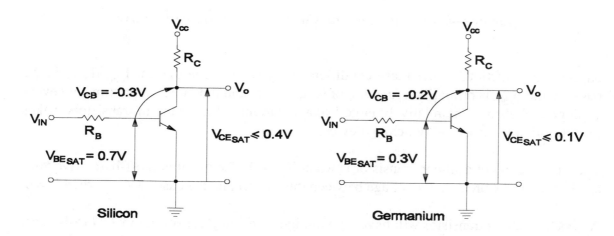

Figure 6-3. Voltage levels of a saturated transistor switch.

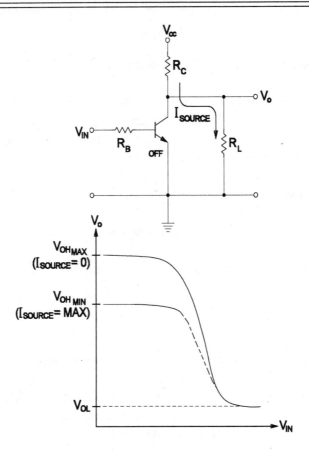

Figure 6-4. Current source transistor switch and its transfer curve.

base-emitter junction. Under these conditions, only the leakage current I_{CBO} flows in the base circuit. I_{CBO} is primarily a function of temperature and it doubles approximately every 10 degrees Celsius. The relatively large leakage currents of germanium transistors makes them unsuitable for use as switching devices.

The cut-off point of a silicon transistor is when $V_{BE} = 0$. The maximum off-base drive voltage is the reverse breakdown voltage between the emitter and the base, BV_{EBO}, when $I_C = 0$.

A good switching transistor will have a good high-frequency cut-off, a low ON resistance, low leakage current, and a higher than average base-emitter reverse breakdown voltage.

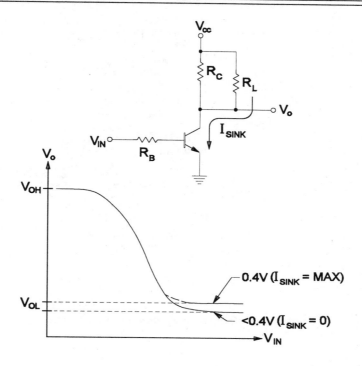

Figure 6-5. *Current sinking transistor switch and its transfer curve.*

Current Sourcing

The transistor switch may be operated as a current source in which it supplies current to a load connected between its collector and ground while the transistor is in cut-off. The output of the inverter switch is in its high state, approximately *Vcc*. A current sourcing transistor is shown in *Figure 6-4*. As the through-load resistor, R_L, decreases, the current flow through it increases, and the output voltage of the inverter falls. The maximum current that the inverter can source is determined by the lowest value of the high level voltage, $V_{OH(min)}$, which the stage following the inverter can tolerate. The maximum allowable load current occurs when the collector voltage has fallen to a specified minimum level; $I_{source}(max) = (Vcc - V_{OH(min)}/R_C)$. The transfer curve of a current sourcing transistor inverter switch is also shown in *Figure 6-4*.

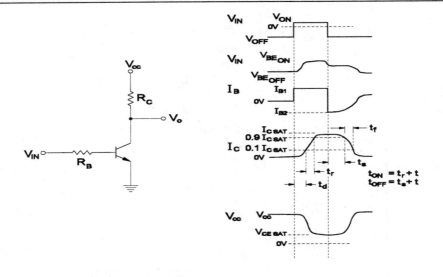

Figure 6-6. *Transistor switch and waveforms.*

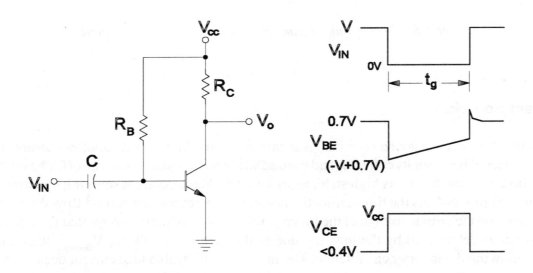

Figure 6-7. *Off-gated switch and waveforms.*

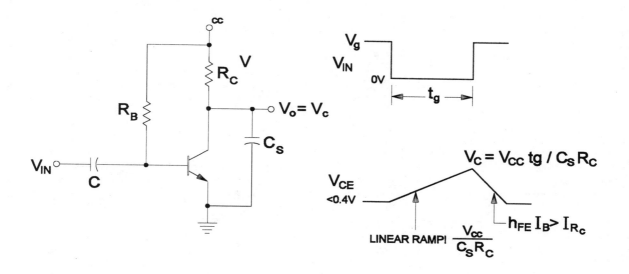

Figure 6-8. Capacitively loaded off-gated switch and waveforms.

Current Sinking

The transistor switch may be operated as a current sink in which it accepts current from a load connected between its collector and *Vcc*. A current sinking transistor is shown in *Figure 6-5*. The current I_{sink} in the load resistor R_L combines sink with the current in the collector resistor, *Rc*, to form the total collector current.

If I_{sink} is increased so that *Ic* is larger than $h_{FE}I_B$, the transistor is no longer saturated. The maximum load current occurs when the total collector current reaches a point such that the available base current is just enough to maintain the transistor in saturation. This point is detected by a rise in collector voltage to a level of 0.4V. Beyond this point, the collector voltage and power dissipation increase rapidly. The transfer curve of a current sinking switch is also shown in *Figure 6-5*.

$I_B = Vin_H - V_{BE(sat)}/R_B$ and $Ic(max) = h_{FE}I_B$, where Vin_H is the peak input voltage.

$I_{RC} = Vcc - V_{OL(max)}/Rc$, where $V_{OL(max)} = V_{CE(sat)}max = 0.4V$.

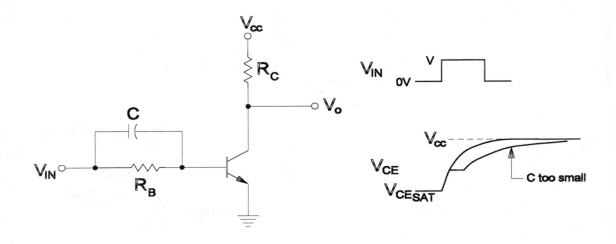

Figure 6-9. Speed-up capacitor switching circuit.

Transistor Switching Speed

The collector current turn-on and turn-off times cannot be instantaneous, even if the input pulse is ideal. The transistor switch and waveforms are shown in *Figure 6-6*. The times, t_d, t_r, t_f and t_s, refer to the transistor delay, rise, fall and storage times, respectively.

Delay Time

The delay time is the time from the instant the turn on voltage is applied until the collector current reaches 0.1 Ic(sat). The transistor begins to conduct when its base-emitter voltage reaches the cut-in voltage, Va, of the base-emitter junction diode. The delay time is the time taken to charge the junction capacitances to Va. A shorter delay time may be designed into the transistor switch by using a smaller R_B. A larger Von for a given Voff will also result in a shorter delay time.

Rise Time

The collector current rise time is the time for the collector current to rise from ten percent to ninety percent of its final value. Collector current flows because there is a diffusion of charge carriers across the field free base region. A charge density gradient is required for diffusion to take place. The mechanism of transistor operation is discussed in the book *Semiconductor Essentials*, by the author. The reader is reminded that $t_r = 2.2RC$; and for the transistor switch, $t_r = 2.2(RaCa + hRcCc)$ where Rc and Cc are the collector resistor and capacitance, respectively, and where $RaCa$ is a proportionality constant.

Fall Time

The fall time is the time it takes for the collector current to fall from ninety percent to ten percent of its final value. During fall time, recombination aids the decay of collector current. At first, recombination is high because there still are a lot of charge carriers present. As the density decreases, the recombination decreases.

Switching Speed

When the transistor operates entirely in the active region, the rise time of the collector current is independent of the base drive voltage and of the base resistor. It is therefore independent of the base current.

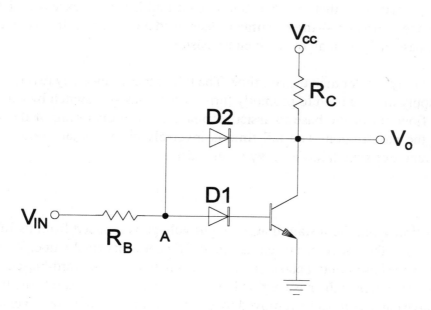

Figure 6-10. Back clamp.

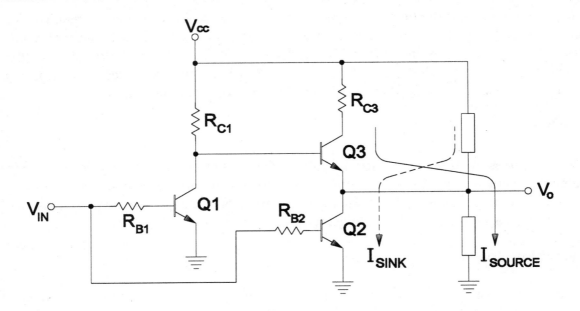

Figure 6-11. *Active pull-up circuit.*

If the transistor is overdriven — that is, the actual base current is well in excess of that just required to saturate the transistor — the rise time is decreased because rise time is inversely proportional to the base current in an overdriven transistor.

The fall time is generally shorter than the rise time. The fall time is reduced by reverse drive. A negative bias supply may be used to promptly turn off the transistor switch by causing a reverse current to flow out of the base to assist recombination in disposing of the stored charge carriers in the base region. The fall time is inversely proportional to the reverse current when the transistor switch has a heavy reverse drive.

Storage Time

Storage time is the time from the instant that the input voltage is reduced from its full ON value to the point where the collector current is ninety percent of its final value. When the transistor is saturated, the base-emitter and collector-base junctions are forward-biased. There is an excess of charge carriers in the base region beyond that required to saturate the transistor. The excess charge carriers must be removed by either recombination or by reverse base drive before the collector current can decay.

Off-Gated Switch

The off-gated switch shown in *Figure 6-7* allows the output to stay at ground potential, until the input gating signal turns off the transistor switch and releases the output from ground potential. The transistor must be saturated prior to the input gate, requiring $R_B < h_{FE}Rc$.

For a silicon device, $V_{BE} = V_{BE}(\text{sat}.7V$ and $V_{CE} = V_{CE}(\text{sat}) < 0.4V$.

When the control gate is applied to the circuit, the input side of the capacitor voltage drops immediately by the step voltage of the control gate. An identical voltage drop appears at the transistor base. The base-emitter voltage becomes negative and the transistor cuts off. The base-emitter diode is no longer conducting and the base is no longer clamped to ground potential. The base voltage rises exponentially towards *Vcc* as the capacitor charges through the base resistor. As long as the capacitor is large enough, the base voltage will not reach the cut-in voltage of the base-emitter diode and the transistor will not turn on before the input gate is terminated.

A linear ramp waveform may be generated by an off-gated switch that has a capacitive load, as shown in *Figure 6-8*. Prior to the input gate, the capacitor is discharged by the saturated transistor. When the transistor is cut off, the capacitor charges toward *Vcc*, with the *time*

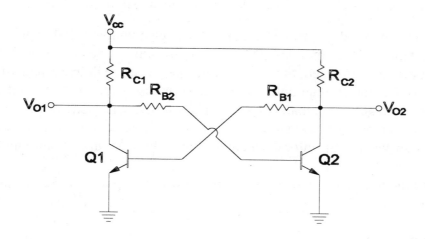

Figure 6-12. Bistable multivibrator circuit.

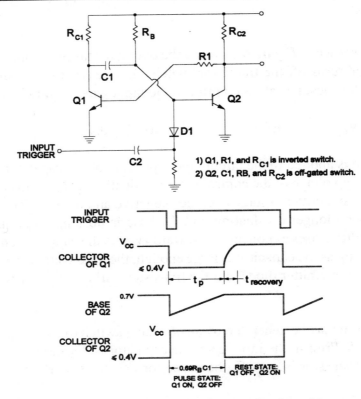

Figure 6-13. Monostable multivibrator circuit and its waveforms.

constant CsRc and an initial slope of Vcc/CsRc. If the gate width, t_g, is less than CsRc/10, the slope is constant, and the amplitude of the ramp, *Vc*, approximately equals $Vcct_g$/CsRc. When the input gate is terminated, the capacitor discharges through the conducting transistor. The transistor cannot saturate until the capacitor is discharged to the point where the collector-emitter voltage is less than 0.4 volts. The transistor remains in its active region with a constant collector current equal to $h_{Fe}I_B$. If the collector current is much larger than the current through *Rc*, the capacitor discharges quickly with an almost constant slope.

The linear charging may be maintained over a much wider amplitude if *Rc* is replaced with a constant current source.

Speed Up Capacitor

The turn-on and turn-off times of a transistor switch may be reduced by placing a capacitor across the base resistor, as shown in *Figure 6-9*. The capacitor bypasses the base resistor during the turn-on and turn-off input voltage steps. The former provides temporary overdrive without deeply saturating the transistor, and the latter provides a temporary increase in reverse current without using a large OFF voltage.

The capacitor should not be larger than necessary to ensure rapid turn-off of the transistor, usually in the 50 to 200 pF range. If the capacitor is too small, the transistor has a fast turn-off for a short time, but slows down as the capacitor charges, as shown in *Figure 6-9*. Integrated circuits do not require speed-up capacitors.

Back Clamp

The switching speed of a transistor can be increased if the transistor is not saturated. This may be accomplished by biasing the transistor into its active region with an emitter resistor, as done in emitter coupled logic; or it can be done by placing a clamp on the base current when the collector falls to a suitable low voltage, generally just above the edge of saturation. The back or Baker clamp circuit is shown in *Figure 6-10*.

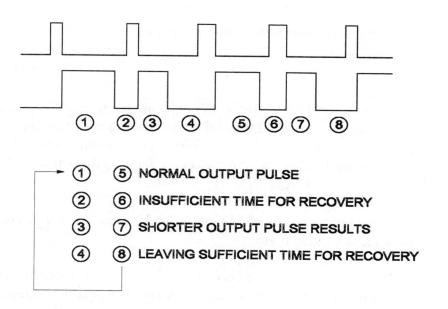

Figure 6-14. Monostable multivibrator waveforms at a higher frequency.

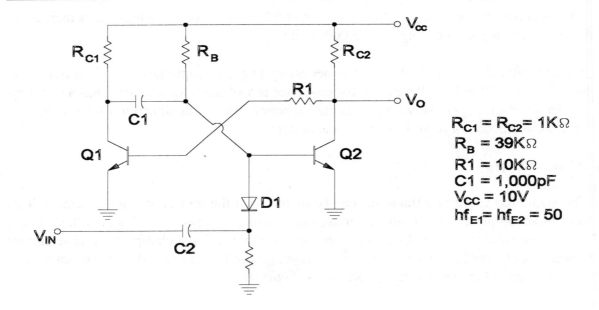

Figure 6-15. Monostable multivibrator circuit for Example 6-1.

$R_{C1} = R_{C2} = 1K\Omega$
$R_B = 39K\Omega$
$R1 = 10K\Omega$
$C1 = 1,000pF$
$V_{CC} = 10V$
$hf_{E1} = hf_{E2} = 50$

When the input voltage is low, the transistor is off or conducting lightly. The collector voltage is higher than the voltage at A, and D2 is off. When the input voltage increases to the point where the collector voltage falls below the voltage at A, D2 turns on, and any excess base current is diverted from the base into the collector circuit. In this situation, $V_A = V_{D1} + V_{BE} = V_{D2} + V_{CE}$, and $V_{CE} = V_{D1} + V_{BE} - V_{D2}$.

If diode D2 and the transistor are of the same type, $V_{D2} = V_{BE}$ and $V_{CE} = V_{D1}$. The collector voltage is kept above the saturation region.

In some integrated circuits, a Schottky diode is fabricated between the base and the collector by arranging the aluminum contact to the base, which also contacts the collector without intervening the *n*-layer. Therefore, $V_{CE}(min) = V_{BE} - V_D = 0.7 - 0.3 = 0.4V$, and the transistor operation is limited to the edge of saturation. This device is called a Schottky transistor.

Active Pull-Up

The standard transistor switch usually sinks current better than it sources current because of its relatively high output impedance, Rc, when the transistor is off. This deficiency is eliminated by the active pull-up switch shown in *Figure 6-11*. In this circuit, Q2 is the normal switch transistor, which can sink current from the load when the input voltage is high. When the input voltage is low and Q2 turns off, Q3 is turned on by the high level voltage on the collector of Q1. This provides a low impedance source which can rapidly charge any capacitive load on the output. The resistor in the collector of Q3 is a low value current limiting resistor.

Bistable Multivibrator

The bistable multivibrator has two stable states. A collector-coupled bistable multivibrator is shown in *Figure 6-12*. It is essentially two inverter transistor switches connected back-to-back. The collector voltage of the conducting transistor is nearly zero volts, while the collector voltage of the nonconducting transistor is approximately Vcc.

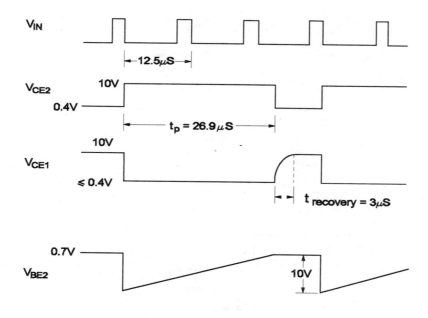

Figure 6-16. Waveforms for Example 6-1.

Both stable states are equally probable. The collector voltage will remain in either state until something disturbs it enough to make it change state. A negative signal coupled to the base of transistor Q1 would decrease its current flow; the collector voltage of transistor Q1 would increase. Transistor Q2 would turn on and its collector voltage would decrease, causing transistor Q1 to turn off. The original negative signal initiates the regenerative process. If the negative signal is large enough, transistor Q1 completely turns off and transistor Q2 saturates.

The stable state achieved by the bistable multivibrator depends upon the nature of the negative signal initiating the change, because both stable states are equally probable.

Monostable Multivibrator

The monostable multivibrator has one stable state. The monostable multivibrator is an off-gated transistor switch connected back-to-back to an inverter transistor switch, as shown in *Figure 6-13*.

When the circuit is in its rest state, transistor Q2 is saturated as long as $R_B < h_{FE}R_{C2}$.

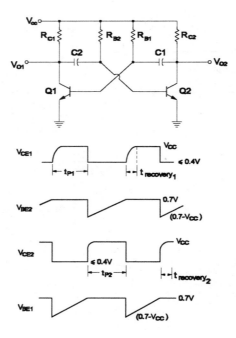

Figure 6-17. *Astable multivibrator circuit and its waveforms.*

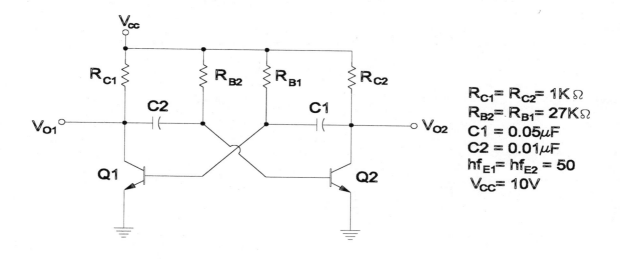

Figure 6-18. Astable multivibrator circuit for Example 6-2.

A negative input trigger pulse is coupled to the base of Q2, initiating the turn-off of transistor Q2. The collector voltage of Q2 increases, causing transistor Q1 to turn on. The decreasing collector voltage of Q1 sustains the turn-off of Q2, because Q1 is coupled by capacitor C1 to the base of Q2. A regenerative snap-action results in which Q2 turns off and Q1 turns on, in a time which is dependent only on the switching speed of the transistors. Transistor Q2 remains off until its base voltage recovers to its cut-in voltage, Va, with a time constant of R_BC1. The waveforms are shown in *Figure 6-13*.

The trigger input capacitor, C2, isolates the base of Q2 from the DC level of the trigger generating circuit. Capacitor C2 also differentiates the input waveform. This makes even a wide pulse useful in initiating an output pulse. Diode D1 allows only negative-going turn-off signals to reach the base of transistor Q2.

The pulse duration is determined by the time it takes the timing capacitor C1 to discharge from Vcc in the rest state to $-Vcc$ in the pulse state. Transistor Q1 is ON and transistor Q2 is OFF in the pulse state. Timing capacitor C1 discharges through R_B and Q1. When the base voltage of Q2 is zero, the capacitor voltage is zero, and transistor Q2 turns on. At this point, the timing capacitor C1 has discharged through half the step of $2Vcc$. Transistor Q2 has been off for 0.69 times the time constant; that is, $t_p = 0.69R_BC1$.

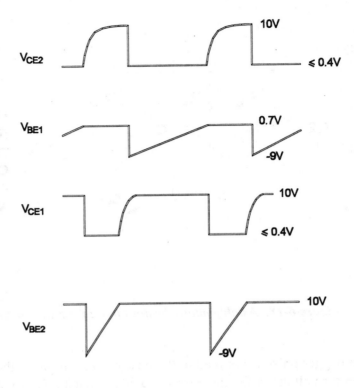

Figure 6-19. Waveforms for Example 6-2.

After transistor Q2 has turned on again, C1 must charge back to its rest state value of *Vcc* through R_{C1} and the base of Q2. The recovery time constant is $R_{C1}C1$, and the recovery time constant is defined as the time necessary for V_{CE} to recover to ninety-five percent of *Vcc* or more; that is, $t_{recovery} > 3CR_{C1}$.

The recovery time is represented by the exponential trailing edge to the collector of the Q1 waveform. Pulse duration errors occur when the monostable multivibrator is triggered during the recovery time because the step at the base of Q2 will not be equal to *Vcc*.

The minimum period of a monostable multivibrator is the sum of its pulse duration and its recovery time; that is, $Tmin = 0.69R_{B}C1 + 3R_{C1}C1$. Its maximum frequency, Fmax, is BC1Fmax = 1/T.

If the frequency of the input trigger is only slightly higher than Fmax, the next trigger pulse falls during the recovery time, and a shorter output pulse results. If there is enough time for the multivibrator to recover, the next output pulse is normal. The following output pulse will again be short, as shown in *Figure 6-14*.

The monostable multivibrator ignores trigger pulses that occur during its pulse state, and during the early part of the recovery time. In this manner, the monostable multivibrator may be used as a frequency divider.

Example 6-1

For the monostable multivibrator shown in *Figure 6-15*, calculate its maximum frequency of operation. What is the duty cycle of the output waveform? Draw the appropriate waveforms when the input pulse train has a frequency of 80 kHz. Assume that for each transistor, $h_{FE} = 50$.

Verify that Q2 saturates. $R_B = 39k < h_{FE}R_{C2} = 50 * 1k$.

Since 39k < 50k, transistor Q2 saturates.

$t_p = 0.69R_BC1 = 0.69 * 39k * 1000 \text{ pF} = 26.9uS$.

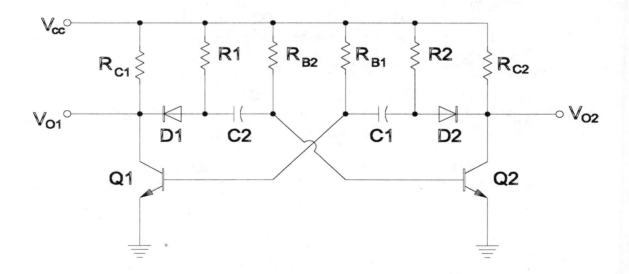

Figure 6-20. An astable circuit with a true squarewave output.

$t_{recovery} > 3R_{C1}C1 = 3 * 1k * 1000 \text{ pF} = 3uS.$

$Tmin = t_p + t_{recovery} = 29.9 \text{ uS}.$

$Fmax = 1/Tmin = 33 \text{ kHz},$ and duty cycle $= t_p/Tmin = 26.9/29.9 = 90\%.$

Amplitude $= Vcc = 10$ volts. The waveforms are shown in *Figure 6-16*.

Astable Multivibrator

The astable multivibrator has no stable state. It will continuously oscillate between its two states; therefore, the astable multivibrator acts like a free-running oscillator. The astable multivibrator consists of two off-gated transistor switches connected back-to-back. The coupling between the two gates is provided by capacitors C1 and C2. If either transistor turns on, the other transistor turns off. The astable multivibrator, like the monostable multivibrator, is a closed-loop, positive feedback, regenerative circuit where the transistors turn on and off with a "snap" action. The astable multivibrator circuit and its waveforms are shown in *Figure 6-17*.

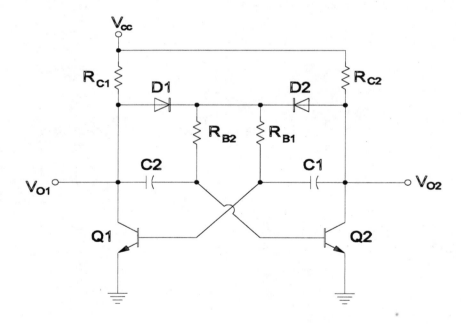

Figure 6-21. A "sure start" astable multivibrator circuit.

The frequency of operation of the astable multivibrator may be calculated as $t_{p1} = 0.69R_{B1}C1$, and $t_{p2} = 0.69R_{B2}C2$.

$$T = t_{p1} + t_{p2} = 0.69(R_{B1}C1 + R_{B2}C2)$$

$$F = 1/T$$

Example 6-2

Calculate the frequency and duty cycle of the output of the astable multivibrator shown in *Figure 6-18*. Draw its waveforms.

Both transistors can saturate if $R_{B1} < h_{FE1}R_{C1}$ and $R_{B2} < h_{FE2}R_{C2}$, which is the case because $R_{B1} = R_{B2} = 27k$ and $h_{FE1}R_{C1} = h_{FE2}R_{C2} = 50k$.

$$t_{p1} = 0.69R_{B1}C1 = 0.69 * 27k * 0.05uF = 930 \text{ uS.}$$

$$t_{p2} = 0.69R_{B2}C2 = 0.69 * 27k * 0.01 \text{ uF} = 186 \text{ uS.}$$

$$T = t_{p1} + t_{p2} = 930 + 186 = 1116 \text{ uS.}$$

$$F = 1/T = 896 \text{ Hz.}$$

$$\text{Duty cycle} = t_{p2}/T = 186/1116 = 0.17.$$

The waveforms are shown in *Figure 6-19*. The basic astable multivibrator has some disadvantages. The slow exponential leading edges of both collector waveforms make the output unsuitable where a true square waveform is required. An astable multivibrator circuit that yields a true squarewave output is shown in *Figure 6-20*.

When power is first applied to an astable multivibrator, both may saturate simultaneously. The circuit will not oscillate until the current in one of the transistors is reduced by external means. A "sure start" astable multivibrator circuit is shown in *Figure 6-21*.

Problems

Problem 6-1. The basic transistor switch has Rc = 600 ohms. What value of h_{FE} should be used at room temperature? What base current will saturate the transistor at room temperature? What is the largest base resistor that can be used if the transistor is to remain saturated at -55 degrees Celsius?

Problem 6-2. Repeat *Problem 6-1* if Rc = 6000 ohms.

Problem 6-3. In an effort to improve the turn-on time of a transistor, the base current is substantially increased. What effect does this have on the turn-off time?

Problem 6-4. In order to decrease the turn-off time of a transistor, a resistor is connected between the base and a negative bias voltage. What effect does this have on the turn-on time?

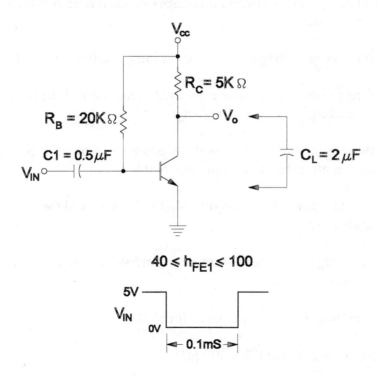

Figure P6-1. Circuit for Problem 6-5.

Problem 6-5. For the off-gated switch shown in *Figure P6-1*, draw the input, base and collector waveforms, under each other showing the correct time relationships and voltage levels. Explain the base waveform. What is the maximum base resistor that can be used in this circuit? A 2 uF capacitor is connected between the collector and ground. What is the slope and the amplitude of the new collector waveform? Sketch the new collector waveform.

Problem 6-6. What is meant by active pull-up? What is the purpose of using it?

Problem 6-7. What is a monostable multivibrator? Draw its waveforms.

Problem 6-8. What basic circuits are used to build a monostable multivibrator?

Problem 6-9. Which transistor is normally conducting in the rest state?

Problem 6-10. What are the approximate voltages on each base and on each collector in the rest state?

Problem 6-11. What is the voltage across the timing capacitor during the rest state?

Problem 6-12. A negative trigger pulse is applied to the input. Will the circuit trigger on its leading or on its trailing edge?

Problem 6-13. What is the voltage across the capacitor and on each base and collector the instant after the circuit has switched?

Problem 6-14. If at this point, Q2 turns off, what voltage would the timing capacitor charge to?

Problem 6-15. At what point does Q2 turn on again? What is the voltage across the timing capacitor?

Problem 6-16. How long has Q2 remained off if:

(a) R_B = 22k and C2 = 500 pF?

(b) R_B = 18k and C2 = 0.01 uF?

Problem 6-17. What is the maximum value allowable for the timing or base resistor of an astable multivibrator if $h_{FE} = 20$ and $R_C = 1.5$ kohm?

Problem 6-18. What is the frequency of oscillation and the duty cycle of an astable multivibrator if $R_{B1} = R_{B2} = 1.5$k, C1 = 0.02 uF and C2 = 0.01 uF?

Chapter Seven

Multivibrators Using Integrated Circuits

Integrated circuit Schmitt triggers and multivibrators are available. These integrated circuits are useful whenever critical shaping or timing is required in a digital circuit.

There are many applications in which a trigger pulse has to be derived from another waveform, a register clear pulse is required, or a small non-critical delay is needed. For these applications, a multivibrator can be easily made from two or three logic gates and one or two discrete components. These circuits are not true multivibrators because they are not regenerative circuits. However, they produce outputs of definite shape or time characteristics when the inputs have certain restrictions. They can usually be converted to regenerative circuits by additional connections or gates.

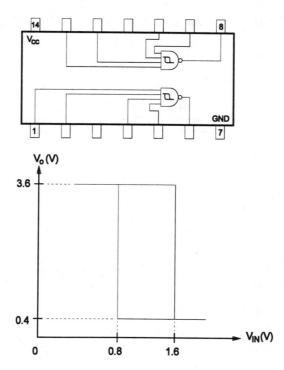

Figure 7-1. The SN7413 dual Schmitt trigger.

Schmitt Trigger

The SN7413 is a dual TTL logic Schmitt trigger. Each circuit operates as a four-input NAND gate. The gate has different input threshold levels for positive and negative going inputs because of the Schmitt action. The hysteresis voltage is 0.8 volts, the UTL is 1.6 volts and the LTL is 0.8 volts. The pin assignment and the transfer curve of the SN7413 are shown in *Figure 7-1*.

An operational amplifier may be used as a Schmitt trigger, as shown in *Figure 7-2*. If the output switches between +Vo and -Vo, the corresponding two voltages on the non-inverting input are +BVo and -BVo respectively, where B = R1/(R1 + R2). The output remains at +Vo as long as Vin is more negative than +BVo. If Vin is more positive than +BVo, the output switches to -Vo. The upper trip level is +BVo and the lower trip level is -BVo. The hysteresis voltage, V_H, is 2BVo. If the input signal is completely above ground, a bias reference voltage, as shown in *Figure 7-3*, may be used to shift the trip levels above ground. When Vo = +Vo, V1 = (1 - B)Vr + BVo, where (1 - B) = R2/(R1 + R2) and Vr is the bias reference voltage. When Vo = -Vo, V2 = (1 - B)Vr - BVo.

Therefore UTL = (1 - B)Vr + BVo and LTL = (1 - B)Vr - BVo, and the hysteresis voltage, V_H = 2BVo.

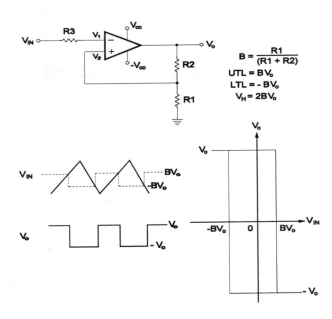

Figure 7-2. Operational amplifier Schmitt trigger without bias reference voltage.

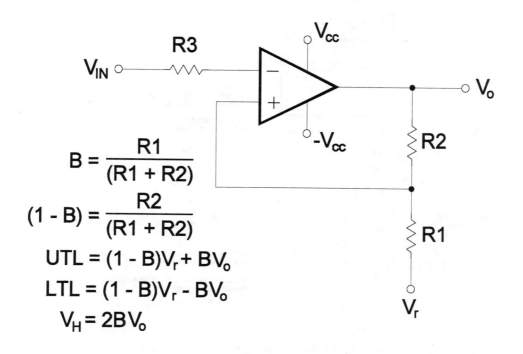

$$B = \frac{R1}{(R1 + R2)}$$

$$(1 - B) = \frac{R2}{(R1 + R2)}$$

$$UTL = (1 - B)V_r + BV_o$$

$$LTL = (1 - B)V_r - BV_o$$

$$V_H = 2BV_o$$

Figure 7-3. Operational amplifier Schmitt trigger with bias reference voltage.

Monostable Multivibrator

The SN74121 is a level-triggered monostable multivibrator. Once it is triggered, it cannot be retriggered until the output pulse is completed, that is, it is not retriggerable. Timing control is by an external resistor and capacitor. Pulse widths from 30 nanoseconds to 40 seconds are possible. The SN74121 monostable multivibrator integrated circuit is shown in *Figure 7-4*.

The SN74122 is a retriggerable monostable multivibrator with a clear input. The SN74122 monostable multivibrator integrated circuit is shown in *Figure 7-5*.

The SN74123 is a dual retriggerable monostable multivibrator with clear inputs. The SN74123 monostable multivibrator integrated circuit is shown in *Figure 7-6*. The retrigger facility simplifies the generation of extremely long output pulses by triggering the input before the output pulse is terminated. The overriding clear pulse permits the output pulse to be terminated at a predetermined time independently of the timing resistor and capacitor.

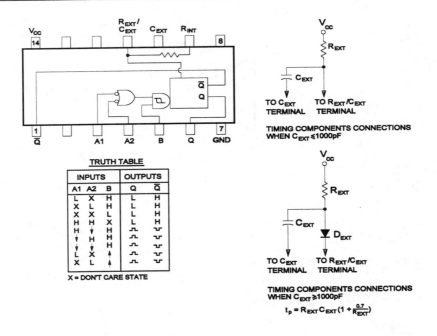

TRUTH TABLE

INPUTS			OUTPUTS	
A1	A2	B	Q	Q̄
L	X	H	L	H
X	L	H	L	H
X	X	L	L	H
H	H	X	L	H
H	↓	H	⊓	⊔
↓	H	H	⊓	⊔
↓	↓	H	⊓	⊔
L	X	↑	⊓	⊔
X	L	↑	⊓	⊔

X = DON'T CARE STATE

TIMING COMPONENTS CONNECTIONS WHEN $C_{EXT} \leqslant 1000pF$

TIMING COMPONENTS CONNECTIONS WHEN $C_{EXT} \geqslant 1000pF$

$$t_p = R_{EXT} C_{EXT} (1 + \tfrac{0.7}{R_{EXT}})$$

Figure 7-4. The SN74121 monostable multivibrator.

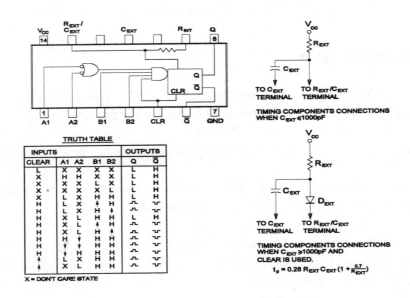

TRUTH TABLE

INPUTS					OUTPUTS	
CLEAR	A1	A2	B1	B2	Q	Q̄
L	X	X	X	X	L	H
X	H	H	X	X	L	H
X	X	X	L	X	L	H
X	X	X	X	L	L	H
X	L	X	H	H	L	H
H	L	X	↑	H	⊓	⊔
H	L	X	H	↑	⊓	⊔
H	X	L	H	H	L	H
H	X	L	↑	H	⊓	⊔
H	X	L	H	↑	⊓	⊔
H	H	↓	H	H	⊓	⊔
H	↓	↓	H	H	⊓	⊔
H	↓	↓	H	H	⊓	⊔
H	L	X	H	H	⊓	⊔
↑	X	L	H	H	⊓	⊔

X = DON'T CARE STATE

TIMING COMPONENTS CONNECTIONS WHEN $C_{EXT} \leqslant 1000pF$

TIMING COMPONENTS CONNECTIONS WHEN $C_{EXT} \geqslant 1000pF$ AND CLEAR IS USED.

$$t_p = 0.28 \, R_{EXT} C_{EXT} (1 + \tfrac{0.7}{R_{EXT}})$$

Figure 7-5. The SN74122 retriggerable monostable multivibrator with a clear input.

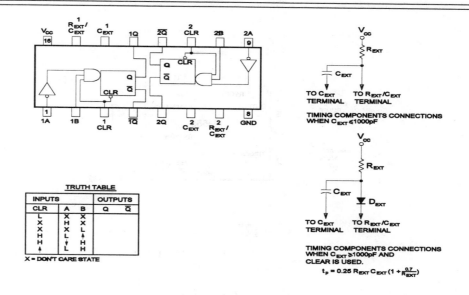

TRUTH TABLE

INPUTS			OUTPUTS	
CLR	A	B	Q	Q̄
L	X	X		
X	H	X		
X	X	L		
H	L	↑		
H	↓	H		
↑	L	H		

X = DON'T CARE STATE

TIMING COMPONENTS CONNECTIONS
WHEN $C_{EXT} \leq 1000pF$

TIMING COMPONENTS CONNECTIONS
WHEN $C_{EXT} \geq 1000pF$ AND
CLEAR IS USED.

$$t_p = 0.25 R_{EXT} C_{EXT} (1 + \frac{0.7}{R_{EXT}})$$

Figure 7-6. The SN74123 dual retriggerable monostable multivibrator with clear inputs.

The LM555 integrated circuit timer can also be used as a monostable multivibrator, as shown in *Figure 7-7*. The external timing capacitor is initially held discharged by a transistor inside the timer integrated circuit. When a negative trigger pulse of less than 1/3 Vcc is applied to the trigger input, pin 2, the capacitor is allowed to charge and the output, pin 3, goes high. The capacitor voltage increases exponentially for a period of t = 1.1R1C1, where R1 and C1 are the timing resistor and capacitor, respectively. When the capacitor voltage reaches 2/3 Vcc, the comparator inside the timer resets the flip-flop, the capacitor discharges and the output, pin 3, goes low. While the output, pin 3, is high, the monostable multivibrator cannot be retriggered. However, the circuit can be reset while the output, pin 3, is high, by applying a negative pulse to the reset terminal, pin 4. The output, pin 3, goes low until another trigger pulse is applied. If the reset function is not being used, the reset terminal, pin 4, should be connected to Vcc to avoid false triggering. The waveforms of the LM555 monostable multivibrator are also shown in *Figure 7-7*. The output pulse width, t_p, is t_p = 1.1R1C1

An operational-amplifier monostable multivibrator is shown in *Figure 7-8*. When the circuit is triggered by an input pulse, it switches to the unstable state for a predetermined period of time. The monostable multivibrator then returns to its stable state until it is triggered by another input pulse. The timing diagram for the operational-amplifier monostable multivibrator is shown in *Figure 7-9*.

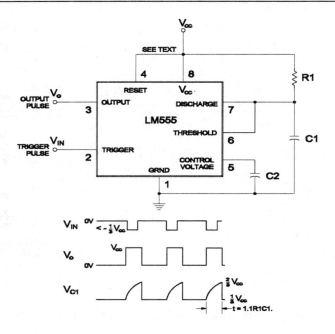

Figure 7-7. *The LM555 as a monostable multivibrator.*

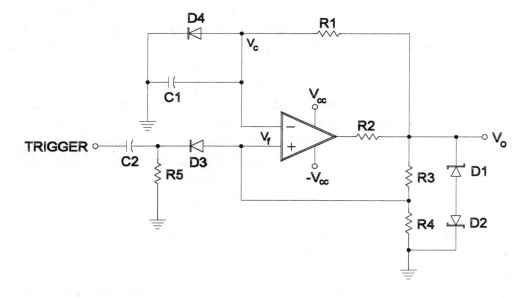

Figure 7-8. *Operational-amplifier monostable multivibrator.*

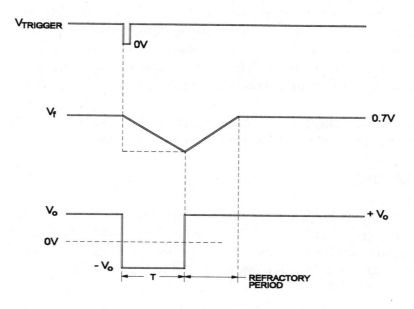

Figure 7-9. Operational-amplifier monostable multivibrator waveforms.

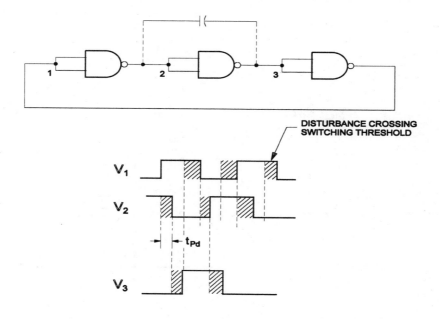

Figure 7-10. Ring oscillator.

The negative-feedback path consists of components R1 and C1. The positive-feedback loop consists of resistors R3 and R4. When a negative-going trigger pulse is applied to differentiator R5 and C2, the output voltage snaps to -Vo, and remains there. Capacitor C1 charges to -Vo, at which point the output snaps high again, ending the output pulse period. Capacitor C1 discharges when Vo is positive. Diode D1 clamps the output voltage to 0.7 volts.

The output pulse period, T, is $T = R1C1 * \ln[(1 + \{0.7V/Vo\})/(1 - R4\{R3 + R4\})]$. If Vo is much larger than 0.7 volts and R3 = R4, then $T = 0.69R1C1$.

Astable Multivibrator

A ring oscillator can be designed using three inverters as shown in *Figure 7-10*. A disturbance crossing the switching threshold of one of the gates travels around the loop generating a self sustaining oscillation of about 10 MHz. The frequency of oscillation can be lowered by inserting a capacitor, which is shown dotted in *Figure 7-10*.

Another version of the ring oscillator using SN7401 open collector gates is shown in *Figure 7-11*. The longer delay of these gates results in a lower frequency of oscillation. If R1 is in the range of 1k-5 kohms and C1 is in the range of 0.05-500 pF, the frequency of oscillation is in the range of 50 KHz to 5 MHz.

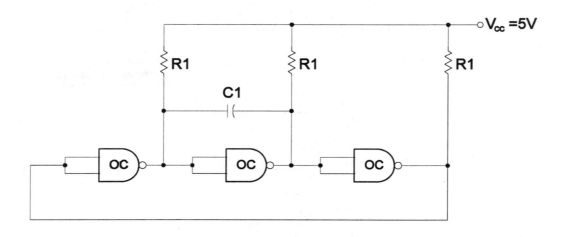

Figure 7-11. Open collector ring oscillator.

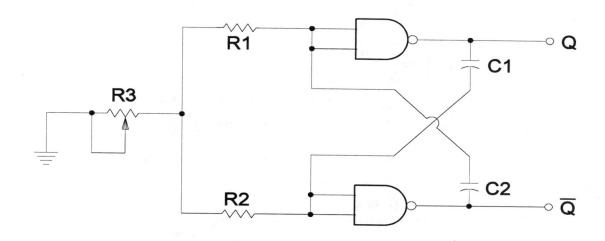

Figure 7-12. Capacitively coupled astable multivibrator.

Lower frequencies of oscillation can be obtained with the astable multivibrator shown in *Figure 7-12*. The gate input sinking resistors R1 and R2 are selected such that the gate input voltage is near its threshold. The charging and discharging of C1 and C2 will switch the input above and below this threshold. The frequency of oscillation, F, is F = 1/2RC, where R = R3 = R1 + R2, R1 = R2, and C1 = C2.

An astable multivibrator may be designed using an operational amplifier, as shown in *Figure 7-13*. The output switches from -Vo to +Vo when the voltage across the capacitor, Vc, falls below +BVo. The capacitor then charges from -BVo towards +Vo. When it reaches +BVo, the output switches back to -Vo. The capacitor voltage now decays towards -Vo; when it reaches -BVo, the output switches back to +Vo, and so on. The time constant in each case is R_tC_t. B = R1/(R1 + R2); and if B = 0.46, then (1 - B)/(1 + B) = 0.54/1.46 = 0.37; and t = R_tC_t, then the period, T, is T = $2R_tC_t$; and the frequency, F, is F = $1/2R_tC_t$.

The LM555 timer integrated circuit may be used as an astable multivibrator, as shown in *Figure 7-14*. The circuit is connected so that it will retrigger itself and cause the capacitor voltage to oscillate between 1/3 Vcc and 2/3 Vcc. The external capacitor charges to 2/3 Vcc through resistor R2. The ratio of resistors R1 and R2 can be adjusted to yield the desired duty cycle. If the charge time is t1 and the discharge time is t2, the total period, T, is T = t1

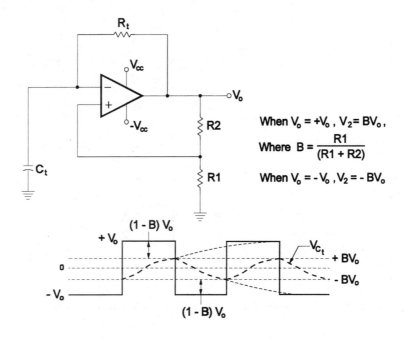

When $V_o = +V_o$, $V_2 = BV_o$,

Where $B = \dfrac{R1}{(R1 + R2)}$

When $V_o = -V_o$, $V_2 = -BV_o$

Figure 7-13. Operational amplifier astable multivibrator.

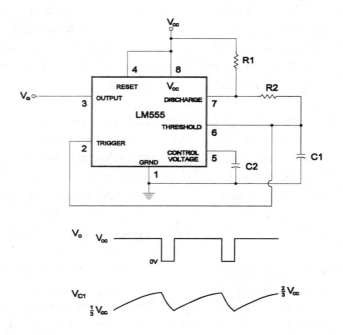

Figure 7-14. The LM555 as an astable multivibrator.

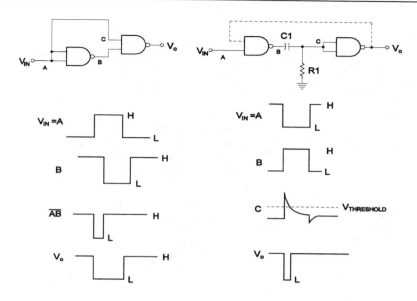

Figure 7-15. *Pulse shortening circuits.*

$+ t2 = 0.693(R1 + R2)C1 + 0.693R2C1 = 0.693(R1 + 2R2)C1$. The frequency of oscillation, F, is $F = 1/T = 1.44/(R1 + 2R2)C1$. The duty cycle, DC, is $DC = R2/(R1 + 2R2)$. To obtain the maximum duty cycle R1 must be as small as possible, but it must also be large enough to limit the discharge current at pin 7 to the maximum rating of 200 mA of the discharge transistor. The maximum value of R1 is $R1 > Vcc/0.2$ The waveforms of the LM555 astable multivibrator are shown in *Figure 7-14*.

Other Circuits

Logic gates may be used as pulse shortening circuits, as shown in *Figure 7-15*. Pulse shortening circuits are useful for flip-flop presetting or clearing purposes.

If several gates are inserted between the input and output of a circuit, as shown in *Figure 7-16*, the gate delays will effectively lengthen the output pulse. The input pulse width must be wider than the desired output pulse width.

An RC delay circuit is shown in *Figure 7-17*. The value of R1 is limited because when the output of gate 1 is low, the current out of the gate 2 input must not develop a voltage across R1. This would lift the resistor voltage drop above the switching threshold. The output pulse width, t_p, is $t_p = 1.2R1C1$.

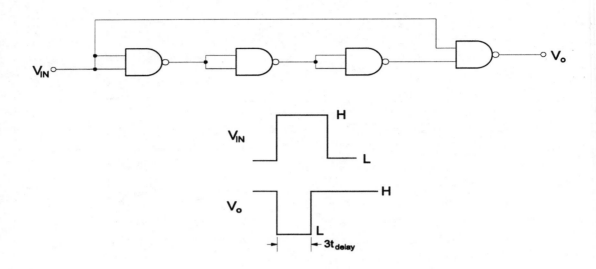

Figure 7-16. Pulse lengthening circuit.

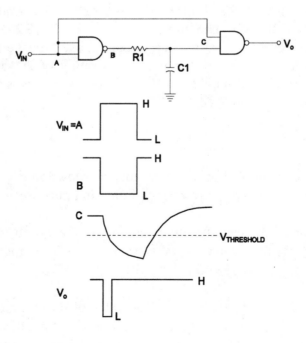

Figure 7-17. RC delay circuit.

Problems

Problem 7-1.   Calculate the trip levels for the Schmitt trigger circuit shown in *Figure P7-1*.

Problem 7-2.   An LM555 timer integrated circuit is used as a monostable multivibrator. If R1 = 10 kohm and C1 = 0.01 uF, calculate the output pulse width.

Problem 7-3.   An operational-amplifier monostable multivibrator has an output voltage of one volt. Calculate the output pulse period if R1 = 10 kohm, R2 = 100 ohm, R3 = R5 = 1 kohm, R4 = 5 kohm, C1 = 0.01 uF and C2 = 0.001 uF.

Problem 7-4.   Repeat *Problem 7-3* if the output voltage is 10 volts and R3 = R4 = 3 kohm.

Problem 7-5.   In a capacitively coupled astable multivibrator, if R1 = R2 = 10 kohm and C1 = C2 = 0.01 uF, calculate its frequency of operation.

Problem 7-6.   In an operational-amplifier astable multivibrator, R_t =1 kohm and C_t =1 uF. Calculate its frequency of operation.

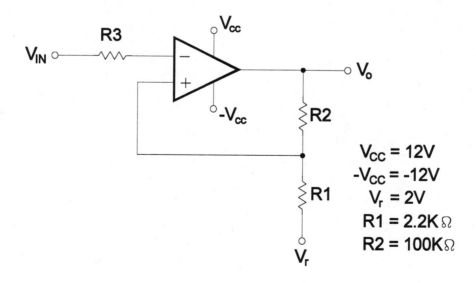

Figure P7-1. Circuit for Problem 7-1.

Problem 7-7. An LM555 integrated circuit timer is used as an astable multivibrator. If Vcc = 15 volts, what is the minimum value of R1 that can be safely used? If R1 = 10 kohm, R2 = 4.7 kohm and capacitor C1 = 0.01 uF, calculate its frequency of operation and the duty cycle of the output waveform.

Problem 7-8. An RC delay circuit has R1 = 22 kohm and C1 = 10 uF. Calculate the output pulse width.

Chapter 8

Blocking Oscillators

The blocking oscillator is a transformer-loaded transistor switch capable of supplying a heavy current pulse to a low impedance load. The blocking oscillator may be used as a monostable multivibrator or as an astable multivibrator. The transformer is connected between the base of the transistor and the collector of the transistor. The blocking oscillator is a regenerative switching circuit.

Inductors and Transformers

An inductor is an electrical component made of several turns of wire that are wound onto a core. The core may be made of air or of a ferromagnetic material. Self-inductance is the property of an electric circuit that opposes any change in current in that circuit. The henry is the basic unit of inductance. An inductor can store energy when a current flows through it. The voltage drop, V_L, across an inductance is $V_L = L(di/dt)$, where L is the inductance and di/dt is the change in current.

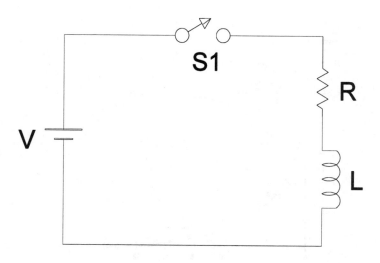

Figure 8-1. Series RL circuit.

A transformer consists of two or more coils that are magnetically coupled. The magnetic field produced in one coil links with the other coil(s) and vice versa, thus transmitting electrical energy from one coil to the other(s). A transformer allows a change in AC voltage and in impedance with only a very small loss in power. A change in the magnetic flux through a coil induces a voltage drop across the coil. The basic transformer equation is V1/V2 = N1/N2, where V1 and V2 are the voltage drops across the primary and secondary windings, respectively, and where N1 and N2 are the number of turns in the primary and secondary windings, respectively. The turns ratio, n, of a transformer is n = N2/N1.

If the secondary voltage of a transformer is higher than its primary voltage, the transformer is a step-up transformer. If the secondary voltage is lower than its primary voltage, the transformer is a step-down transformer.

Shaping with an RL Circuit

All wires have some electrical resistance. Therefore the practical inductor is a series circuit consisting of inductance and resistance. A series RL circuit with a voltage source is shown in *Figure 8-1*. When the switch, S1, is closed, the voltage drop across the resistance of the coil, R, and the voltage drop across the inductance, L, of the coil must equal the applied voltage, V. The voltage drop across the inductance results from the counter emf that is generated in the inductance by the rising current. $V = iR + L(di/dt)$.

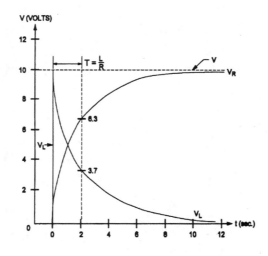

Figure 8-2. Instantaneous voltage drops of a series RL circuit.

When the switch in the DC circuit of *Figure 8-1* is closed, it takes time for the current to rise to its steady-state value. The time constant of an LR circuit is the time it would take the current to rise to its steady-state value if it were to continue to rise at its initial rate of change for the whole time interval. The time constant, T, for an RL circuit is T = L/R.

The resistance of the inductor prevents the instantaneous current from continuing to rise at its initial rate. It rises exponentially as shown in *Figure 8-2*. The time constant is redefined as the time it takes the instantaneous current of an RL circuit to reach sixty-three percent of its steady-state value. The inductor acts like an open circuit at the initial instant that power is applied to the RL circuit. The inductor acts like a short circuit several time constants after power is applied to the RL circuit.

Blocking Oscillator Circuits

The blocking oscillator is used to produce short, high-power, or high-current pulses. The blocking oscillator consumes little power in between pulses; therefore, transistors dissipating less power may be used in blocking oscillator circuits.

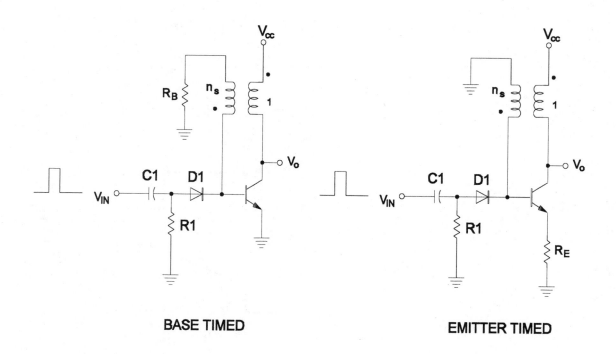

Figure 8-3. Blocking oscillator circuits.

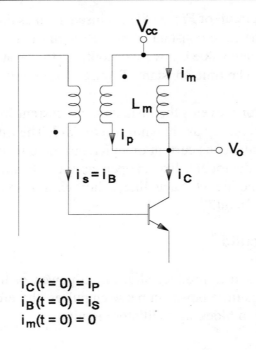

$$i_C(t = 0) = i_P$$
$$i_B(t = 0) = i_S$$
$$i_m(t = 0) = 0$$

Figure 8-4. Blocking oscillator coupling transformer.

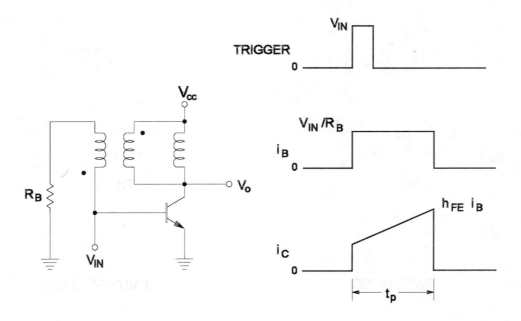

Figure 8-5. Base timed blocking oscillator circuit.

In a monostable blocking oscillator, the transistor has no base drive. The transistor is therefore in cut-off. A positive pulse will trigger the blocking oscillator circuits shown in *Figure 8-3*. Components R1, C1 and D1 condition the positive input pulse. The collector current increases and the collector voltage decreases because of the inductance of the coupling transformer. The base current increases because of the phase reversal connections of the coupling transformer. The increase of the base current causes the collector current to increase. This regenerative action causes the transistor to saturate. As the collector current increases, the transistor can no longer remain in saturation. The collector voltage increases causing the base voltage to decrease and the resulting regenerative action turns the transistor off.

The blocking oscillator transformer can be considered as an ideal turns ratio in parallel with its magnetizing inductance, as shown in *Figure 8-4*. The ratio between the primary and secondary windings is $1:n_s$, where n_s lies in the range 0.2-1. When the transistor is triggered on at $t = 0$, there is no current in the magnetizing inductance, but there is current in the ideal transformer. The magnetizing current is a function of time: $i_m(t) = V_p(t)/L_m$.

The base timed blocking oscillator is shown in *Figure 8-5*. The transistor saturates heavily when it is turned on because the base current is higher than the collector current. The voltage stepdown from primary to secondary causes the base current to be higher than the collector current. $V_p = Vcc$ and $V_s = n_s Vcc$ from which $i_B = n_s Vcc/R_B = $ constant. The collector current increases as the magnetizing current increases until $i_C/i_B = h_{FE}$, at which point the transistor comes out of saturation. The collector-emitter voltage tends to increase and the regenerative action cuts the transistor off. The time t_p is dependent on h_{FE}. This circuit is not used where the stability of the pulse width is important.

The emitter timed blocking oscillator circuit is shown in *Figure 8-6*. This circuit operates in a similar manner to the base timed circuit. The emitter current, rather than the base current, remains constant because the base and emitter voltages are constant. The transistor saturates at first turn-on.

$V_p = Vcc/(1 + n_s)$, and $V_s = n_s Vcc/(1 + n_s)$, and $i_E = i_C + i_B = i_m + i_p + i_s$.

As i_m increases with time during the current pulse, $(i_p + i_s)$ must decrease. When $i_C/i_B = h_{FE}$, the transistor comes out of saturation. The collector-emitter voltage tends to increase and the transistor is regeneratively turned off. If the transistor has a high gain or h_{FE}: $t_p = Li_E/V_p$, where $i_E = V_E/R_E$.

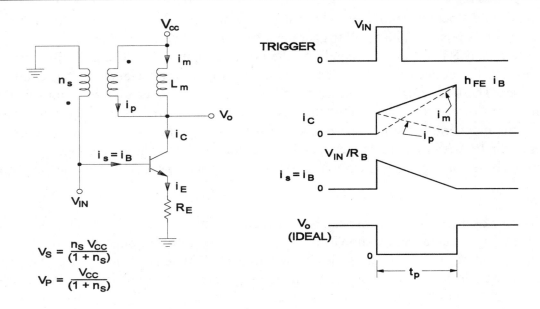

Figure 8-6. *Emitter timed blocking oscillator circuit.*

If a tertiary load winding is provided in which the primary to tertiary turns ratio is $1:n_L$, the current reflected from this winding is $n_L i_L$. When the transistor is cut off, $i_m = i_C - n_L i_L = i_E - n_L i_L$, and $t_p = i_m L_m / V_p = L_m(i_E - n_L i_L)/V_p$.

The pulse width is almost independent of h_{FE} and it is therefore more stable with changes in temperature and transistor parameters. The emitter timed blocking oscillator is more commonly used because its output pulse width is more stable than that of the base timed blocking oscillator.

Example 8-1

Analyze the blocking oscillator circuit shown in *Figure 8-7*, with and without a load.

WITHOUT LOAD:

$V_S = V_E$ and when the transistor is saturated:

$V_p + V_s = Vcc = V_p(1 + n_s)$

$V_p = 9/1.5 = 6$ volts and $V_s = 3$ volts.

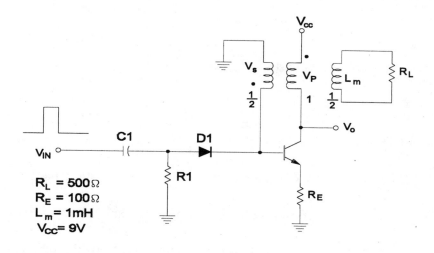

Figure 8-7. *Blocking oscillator circuit for Example 8-1.*

$R_L = 500\,\Omega$
$R_E = 100\,\Omega$
$L_m = 1\,mH$
$V_{CC} = 9V$

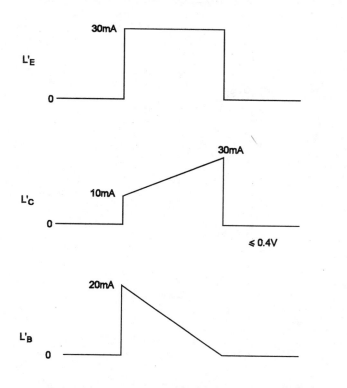

Figure 8-8. *Waveforms for Example 8-1.*

At t = 0, $i_E = V_E/R_E$ = 30 mA, $i_B = V_s/(1 + n_s)R_E$ = 20 mA, $i_C = i_E - i_B$ = 10 mA

and i_m = 0 mA.

At t = t_p, $i_B = i_s$ = 0 mA, i_p = 0 mA and $i_m = i_E$ = 30 mA

$t_p = L_m i_E/V_p$ = 0.001 x 0.03/6 = 5 uS.

WITH 500 OHM LOAD:

$i_L = V_L/R_L$ = 3/500 = 6 mA, $n_L i_L$ = 0.5 x 6 = 3 mA, $i_m = (i_E - i_L)$ = (30 - 3) = 27 mA,

and $t_p = L_m i_m/V_p$ = 0.001 x 0.027/6 = 4.5 uS. The waveforms are shown in *Figure 8-8*.

If the emitter is biased such that it cannot be permanently kept off, the result is an astable blocking oscillator. An astable blocking oscillator is not possible because a transformer cannot transmit a DC level change. A practical astable blocking oscillator is shown in *Figure 8-9*.

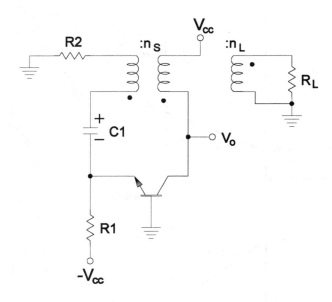

Figure 8-9. *Astable blocking oscillator circuit.*

A capacitor, C1, has been added in series with the primary of the transformer and a bias resistor, R1, has been added to the emitter circuit to bias the transistor "on" in steady-state conditions. The load may be connected to the circuit through a separate transformer winding if the load voltage must be adjusted during the pulse, by selecting the turns ratio n_L.

Problems

Problem 8-1. What is V_p and V_s in the base timed blocking oscillator circuit shown in *Figure P8-1*?

Problem 8-2. Calculate i_B, i_C and i_m when the transistor in *Figure P8-1* turns on. Is the transistor saturated?

Problem 8-3. At what rate does the collector current in *Figure P8-1* increase after turn on?

Problem 8-4. At what point does the transistor in *Figure P8-1* turn off? Calculate the output pulse width.

Problem 8-5. What are the disadvantages of the base timed circuit? How is it affected by temperature or transistor replacement?

Problem 8-6. Calculate the voltages across each winding of the emitter timed blocking oscillator circuit shown in *Figure P8-2*.

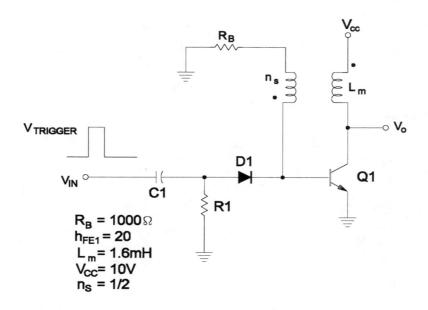

$R_B = 1000\,\Omega$
$h_{FE1} = 20$
$L_m = 1.6mH$
$V_{cc} = 10V$
$n_S = 1/2$

Figure P8-1. *Base timed blocking oscillator for problem set.*

Problem 8-7. Calculate the emitter current of *Figure P8-2*. Does it vary during the current pulse?

Problem 8-8. If the load winding of *Figure P8-2* has no load, calculate i_B, i_p, i_m, i_C and i_E at the end of the current pulse.

Problem 8-9. Calculate the output pulse width of the circuit shown in *Figure P8-2*.

Problem 8-10. How is the pulse width affected if a load resistor of 200 ohms is placed across the load winding of *Figure P8-2*? Calculate the pulse width.

Problem 8-11. What resistor placed directly across the primary of the circuit shown in *Figure P8-2* would produce the same result of *Problem 8-10*?

Problem 8-12. What is the cause of the backswing voltage found in the circuit shown in *Figure P8-2*? What can happen if the backswing is left undamped?

Problem 8-13. How can the backswing be suppressed without affecting the pulse width?

Problem 8-14. How is the maximum output frequency affected by the method used in *Problem 8-13*?

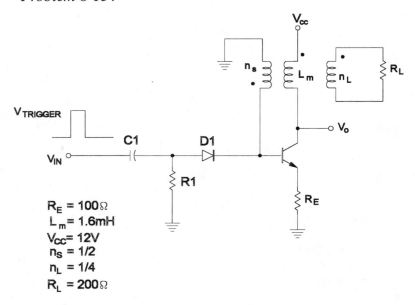

$R_E = 100\Omega$
$L_m = 1.6mH$
$V_{cc} = 12V$
$n_S = 1/2$
$n_L = 1/4$
$R_L = 200\Omega$

Figure P8-2. Emitter timed blocking oscillator for problem set.

Chapter 9

Negative-Resistance Devices and Oscillators

A negative-resistance device has a decreasing current for an increasing voltage in some part of its VI-characteristic curve. Very simple bistable, monostable and astable multivibrator circuits may be made using these devices. Usually only one negative-resistance device in combination with a simple RL or RC circuit is required.

There are two-terminal, three-terminal and four-terminal negative-resistance devices. The negative resistance of two-terminal devices cannot be changed. The tunnel diode, four-layer diode, silicon unilateral switch (SUS), silicon bilateral switch (SBS), and diac are two-terminal negative-resistance devices. The three-terminal and four-terminal negative-resistance

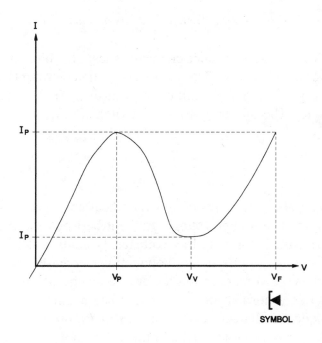

SYMBOL

Figure 9-1*. Tunnel diode VI-characteristic.*

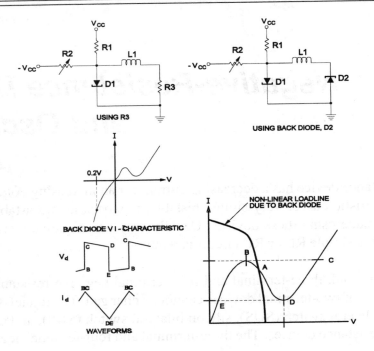

Figure 9-2. Tunnel diode astable oscillator.

devices have variable negative-resistance characteristics. The third terminal is used to change its negative-resistance characteristic. The silicon controlled rectifier (SCR), triac, unijunction transistor (UJT), and programmable unijunction transistor (PUT) are three-terminal negative-resistance devices. The silicon controlled switch (SCS) is a four-terminal negative-resistance device.

Tunnel Diode

The tunnel diode is a PN junction device in which heavy impurity doping results in a very narrow depletion layer. The abrupt voltage gradient between the P and N regions allows conduction across the junction at very low voltages, typically 0.1 volt or less, due to quantum mechanical tunneling. In this mode, conduction decreases as the voltage across the device increases. When the voltage across the device reaches the normal cut-in voltage, Va, the device behaves like a normal junction diode. The normal cut-in voltage is about 0.5 volts. The tunnel diode VI-characteristic is shown in *Figure 9-1*. The negative resistance region is between the peak point (V_p, I_p) and the valley point (V_v, I_v). The tunnel diode is voltage controllable or voltage stable.

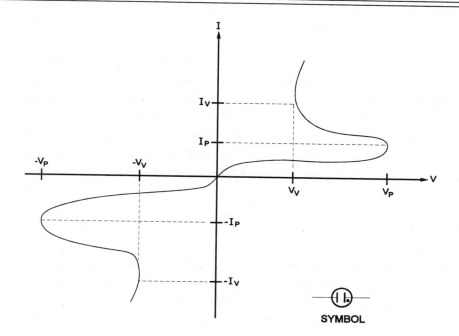

Figure 9-3. Neon tube VI-characteristic.

The tunneling phenomenon takes place at the speed of light. The switching speed of the device is limited by the capacitance of the diode, the circuit and any series lead inductance. Switching times of less than one nanosecond are relatively easy to obtain and switching times of tens of picoseconds are possible. The weak points of a tunnel diode are its low signal swing, and being only a two-terminal device, it is difficult to fabricate it in an integrated circuit.

The tunnel diode may be used in the astable oscillator circuit shown in *Figure 9-2*. It is critical that R3 << R1 and that R3 << R2. The resistor R3 may be replaced by a back diode which is a tunnel diode with a low peak current and operated in the reverse direction. Therefore it conducts with a "diode drop" of 0.2 volts or less. When the tunnel diode switches to its high voltage state, (region B to C) the voltage across the inductor decreases and the current through it decays exponentially. As the operating point moves down the VI-characteristic from C to the valley point D, the tunnel diode returns to its low voltage state E. In region E to B the current and voltage increase exponentially until the peak point is reached again. The circuit waveforms are shown in *Figure 9-2*.

Neon Tube

The neon tube is the oldest negative-resistance device. The tube is filled with neon or other gas. The gas and the pressure with which it is held in the tube determine the breakdown voltage V_p and the sustaining voltage V_v. The breakdown voltage is in the range of 70 to 90 volts and the sustaining voltage is in the range of 55 to 65 volts. The VI-characteristic of the neon tube is shown in *Figure 9-3*. The neon tube is inexpensive, but it switches slowly and the negative-resistance characteristic is unstable and varies from bulb to bulb. The neon tube used to be used as an indicator or read-out device. The neon tube is a current controlled device.

Shockley Diode

The Shockley diode is a four-layer, three-junction current controlled device. Its VI-characteristic is shown in *Figure 9-4*. The Shockley diode may be used as a multivibrator, as shown in *Figure 9-5*. The complementary transistor pair yields a two state circuit in which the circuit current is either very small or potentially very large. $I_E = I_{CO1} + I_{CO2}/(1 - [a1 + a2])$, where a1 and a2 are the transistor cut-in voltages. Since a1 and a2 are very small, $I_E = I_{CO1} + I_{CO2}$.

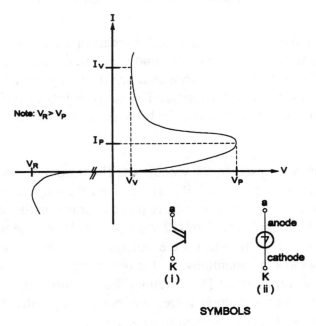

Figure 9-4. Shockley diode VI-characteristic.

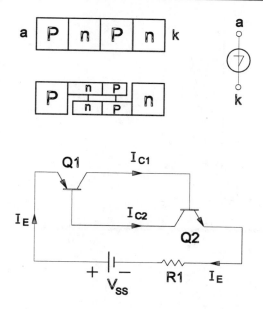

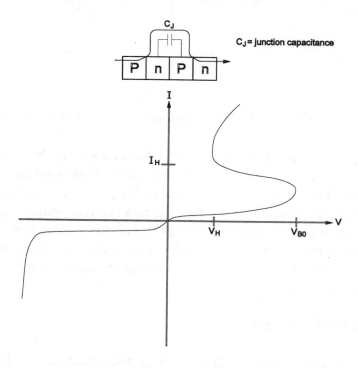

Figure 9-5. *Shockley diode multivibrator.*

C_J = junction capacitance

Figure 9-6. *Shockley diode rate-effect defect.*

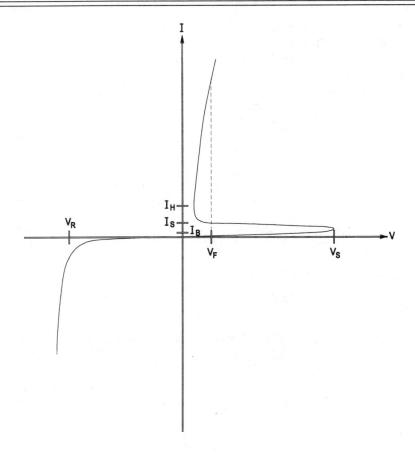

Figure 9-7. Silicon unilateral switch VI-characteristic.

The Shockley diode has a rate-effect defect. Normally, only the middle junction of the device is reversed-biased. Therefore it has the normal reverse-biased junction capacitance which must be charged to the voltage across the junction as illustrated in *Figure 9-6*. A larger charging current will flow in response to a fast rise voltage step applied between the anode and the cathode than when the rate of rise of this voltage is slow. Consequently, (a1 + a2) approaches unity at a lower anode voltage for a fast rise voltage step than for a slow rise voltage step. The device may break over whenever anode voltage is applied to it, if the circuit is not designed properly.

Silicon Unilateral Switch

The silicon unilateral switch (SUS) is an integrated circuit. Its VI-characteristic is shown in *Figure 9-7*. The SUS symbol and equivalent circuit are shown in *Figure 9-8*.

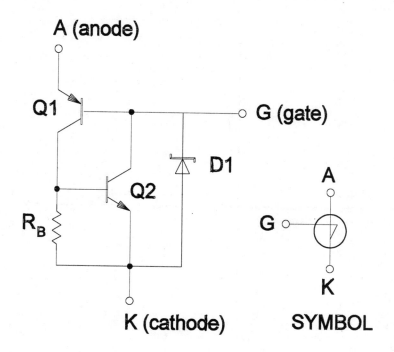

Figure 9-8. Silicon unilateral switch equivalent circuit.

When the voltage applied through a resistor is low, the device is in the low current state. When the voltage applied exceeds the zener voltage, the zener diode allows current to flow to transistor Q1. The higher the applied anode voltage becomes, the more heavily forward-biased Q1 becomes and the more base current transistor Q2 receives. Transistor Q2 conducts and supplies Q1 with more base current. As the base currents increase, the voltage drop across each transistor decreases. The zener current drops until the zener diode turns off. Both transistors saturate and the SUS is in its high current state. The zener diode provides a more precisely controlled break-over point. The gate terminal, G, allows a zener diode with a lower zener voltage to be connected between terminals G and K. The break-over voltage can also be reduced by increasing the current in the off state by connecting a resistor between terminals A and G and a second resistor between terminals G and K to form a voltage divider supplying terminal G with the proper voltage.

The SUS is reasonably fast and can discharge a 0.01 uF capacitor in less than one microsecond. However, the device turn-off time is slower, typically less than 25 microseconds.

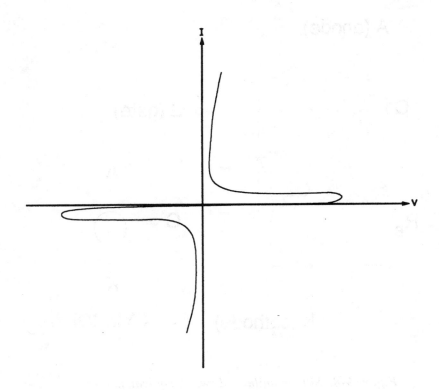

Figure 9-9. Silicon bilateral switch VI-characteristic.

Silicon Bilateral Switch

The silicon bilateral switch (SBS) is essentially two silicon unilateral switch devices connected back-to-back. The VI-characteristic of a SBS is symmetrical as shown in *Figure 9-9*. The symbol and the equivalent circuit of a SBS device is shown in *Figure 9-10*. The SBS is useful in circuits that operate on AC because the device can be made to trigger on the positive and the negative half cycles of the input signal.

Diac

The diac is a three-layer, two-junction device with the VI-characteristic shown in *Figure 9-11*. The diac is analogous to the symmetrical transistor, which may be operated in its avalanche mode.

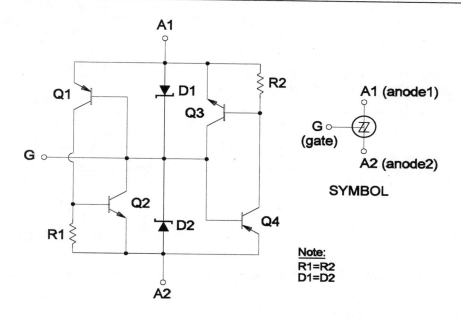

Figure 9-10. *Silicon bilateral switch equivalent circuit.*

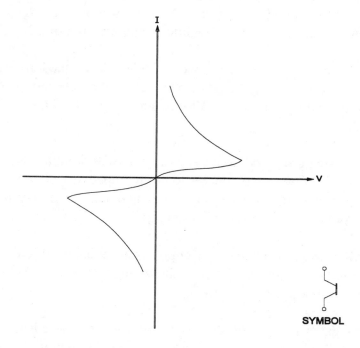

Figure 9-11. *Diac VI-characteristic.*

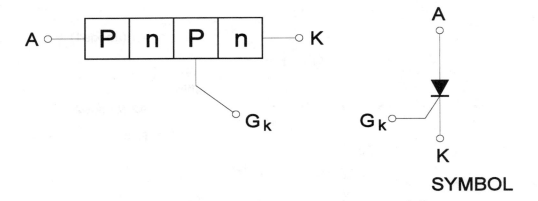

Figure 9-12. Silicon controlled rectifier construction.

Silicon Controlled Rectifier

The silicon controlled rectifier (SCR) is a four-layer device with an additional terminal, the gate terminal, brought out from the internal P-layer, as shown in *Figure 9-12*. A positive pulse on the gate terminal triggers the SCR into the conduction or ON state.

The device breaks over as (a1 + a2) approaches unity. When the current required for this condition is partially provided by the positive input on the gate terminal, (G_k), the break-over voltage is temporarily lowered. The VI-characteristic of the SCR is shown in *Figure 9-13*.

A low power trigger signal may be used to switch the SCR from a high impedance state in which it conducts only a few microamperes to a low impedance state in which it may conduct many amperes. The SCR is prone to rate-effect and therefore measures must be taken to prevent the SCR from being turned on by a too rapid rise of the anode voltage.

The silicon controlled switch (SCS) is a lower powered device in which a second gate terminal, (G_a), is connected to the internal layer as shown in *Figure 9-14*. A negative pulse on G_a has the same effect as a positive pulse on G_k.

In both devices, the gate usually loses control as soon as the device is turned on. Turn-off is achieved by externally reducing the load current to a value below the holding current. A reverse polarity signal on either gate terminal does not turn the device off.

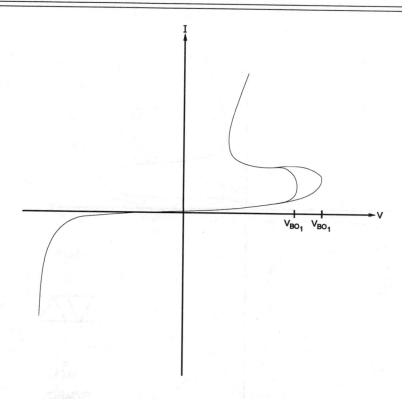

Figure 9-13. Silicon controlled rectifier VI-characteristic.

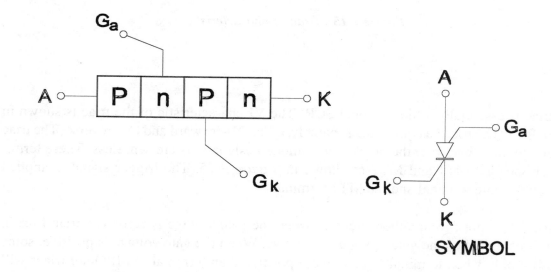

Figure 9-14. Silicon controlled switch construction.

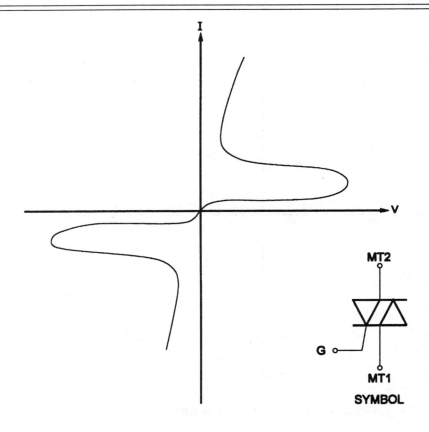

Figure 9-15. *Triac VI-characteristic.*

Triac

The triac is essentially a bidirectional SCR. The VI-characteristic of the triac is shown in *Figure 9-15*. The triac has three states, namely, OFF, ON-forward and ON-reverse. The triac is bidirectional; therefore, the anode and cathode designations are senseless. These terminals are labelled MT1 and MT2 as shown in *Figure 9-15*. The trigger signal is applied between the gate terminal and the MT1 terminal.

The triac does not fire in either direction when the gate voltage is zero. The triac fires in either direction when the gate voltage is negative. When the gate voltage is positive, some triacs fire only when terminal MT2 is more positive than terminal MT1. Other triacs will fire in either direction when MT2 is more positive than MT1.

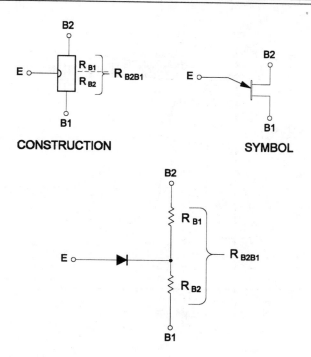

Figure 9-16. Unijunction transistor equivalent circuit.

The triac can be used to switch large currents when a small gate voltage is applied to it. The triac can be used to switch alternating current because the triac has a symmetrical VI-characteristic, as shown in *Figure 9-15*.

Unijunction Transistor

The unijunction transistor (UJT) is a bar of N-type silicon with a P-type alloy junction alloyed to its middle, as shown in *Figure 9-16*. Ohmic contacts are made to each end of the bar. The lower ohmic contact, B1, is usually grounded. A model of the UJT is shown in *Figure 9-16*.

The UJT has the characteristics of a resistor when there is no emitter current. Very little emitter current flows until the emitter voltage, V_{EB1}, is raised sufficiently to forward-bias the PN-junction. This occurs when the lower edge of the emitter becomes more positive than the bar under the emitter. The emitter voltage is called the peak-point voltage and it is denoted V_p.

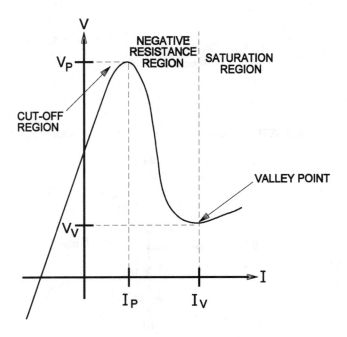

Figure 9-17. Unijunction transistor VI-characteristic.

$V_p = nV_{B2B1} + Va$, where $n = R_{B1}/R_{B2B1}$ and is called the intrinsic stand-off ratio. The intrinsic stand-off ratio is usually in the range of 0.47 - 0.75. The diode cut-in voltage, Va, is about 0.7 volts.

Only leakage current flows when the emitter voltage is less than nV_{B2B1}. As the emitter voltage is increased, emitter current starts to flow. The increase of charge carriers in the R_{B1} section of the silicon bar increases its conductivity progressively so that the voltage across it decreases as the current increases. The UJT therefore has a negative resistance characteristic, as shown in *Figure 9-17*. As the valley point is reached, the normal junction diode characteristic predominates, and the voltage increases slightly as the current is increased further.

A unijunction transistor relaxation oscillator is shown in *Figure 9-18*. The load line must cut the negative slope portion of the UJT VI-characteristic. Therefore, $(V_{EE} - V_p)/I_p > R1 > (V_{EE} - V_v)/I_v$. This poses no problems as long as very long or very short periods are contemplated. The capacitor C1 charges until $V_{C1} = nV_{B2B1} + Va$.

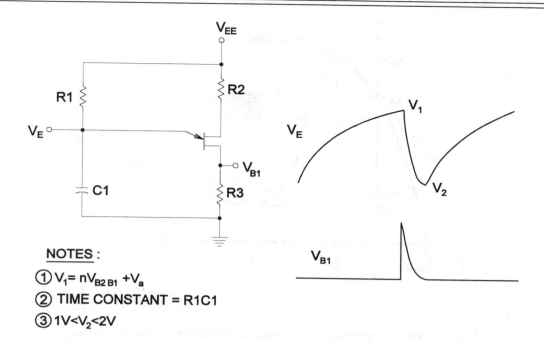

NOTES :

① $V_1 = nV_{B2\,B1} + V_a$

② TIME CONSTANT = R1C1

③ $1V < V_2 < 2V$

Figure 9-18. Unijunction transistor relaxation oscillator.

The UJT then fires and discharges the capacitor until the emitter voltage decays to the valley point. The UJT turns off at this point. The unijunction transistor relaxation oscillator waveforms are shown in *Figure 9-18*.

Programmable Unijunction Transistor

The programmable unijunction transistor (PUT) is a four-layer device and it is more similar to the silicon controlled switch than to the unijunction transistor. Only the anode gate of the SCS is brought out. The SCS will fire when the anode voltage just exceeds the gate voltage, just like a silicon unilateral switch. The peak voltage is programmable by external circuit components and it therefore does not depend on the uncertain intrinsic stand-off ratio, n. In all other respects the PUT anode corresponds to the UJT emitter, and the PUT cathode corresponds to the UJT base 1, B1. The PUT and UJT may be used interchangeably in UJT circuits. The programmable unijunction transistor VI-characteristic is shown in *Figure 9-19* and its transistor equivalent is shown in *Figure 9-20*.

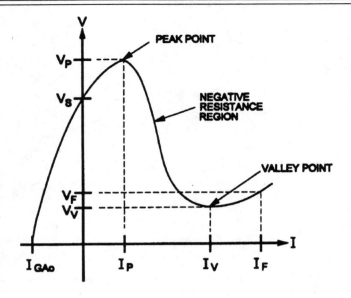

Figure 9-19. Programmable unijunction transistor VI-characteristic.

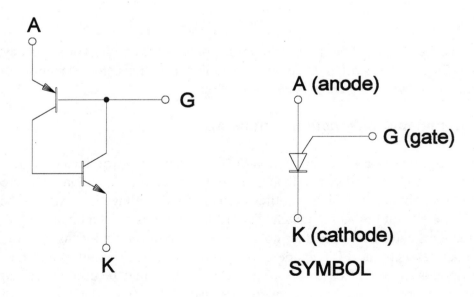

Figure 9-20. Programmable unijunction transistor equivalent circuit.

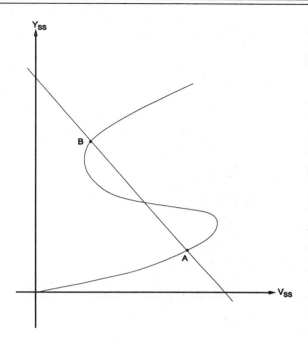

Figure 9-21. Bistable mode of operation load line.

Bistable, Monostable and Astable Modes

Negative resistance devices may be operated in the bistable, in the monostable or in the astable mode. Trigger circuits can therefore be made using a negative resistance device in combination with simple RC or RL circuits.

Bistable operation is obtained when the load line cuts the device VI-characteristic in each positive slope segment, as shown in *Figure 9-21*. Points A and B are the two stable operating points of the bistable circuit. If either X_{ss} or Y_{ss} is momentarily increased, the device triggers from state A to state B. A negative trigger returns the device to state A. In some devices, such as the SCR, the trigger is applied to the gate.

Monostable operation is obtained when the load line cuts the device VI-characteristic only once on a positive slope segment, as shown in *Figure 9-22*. Triggering is achieved as in the bistable mode of operation. In the monostable mode of operation an energy storage component must be provided to hold the device in its second state for a prescribed length of time.

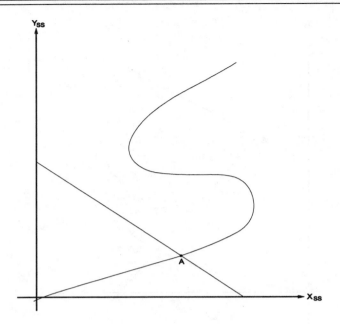

Figure 9-22. Monostable mode of operation load line.

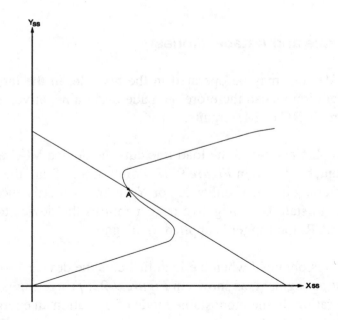

Figure 9-23. Astable mode of operation load line.

In the astable mode of operation, the load line intersects the device VI-characteristic only on its negative slope segment, as shown in *Figure 9-23*.

Negative-Resistance Oscillators

A negative-resistance effect in a circuit means that an increase in the voltage across a circuit is accompanied by a decrease in the current through the circuit. When power is supplied to an oscillator, regenerative feedback introduces a negative resistance. The negative resistance becomes a source of power that overcomes the losses in an oscillator circuit.

The positive-feedback oscillator is built from active circuit components; that is, components that provide gain. The negative-resistance oscillator is built from passive circuit components; that is, components that do not provide gain. Negative-resistance or relaxation oscillators use negative-resistance devices that pass little or no current below the voltage threshold and that pass a relatively large current above the voltage threshold.

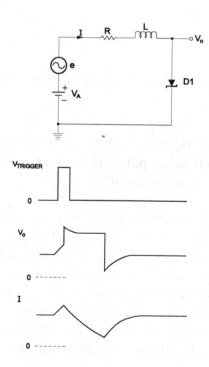

Figure 9-24. Tunnel-diode monostable multivibrator and waveforms.

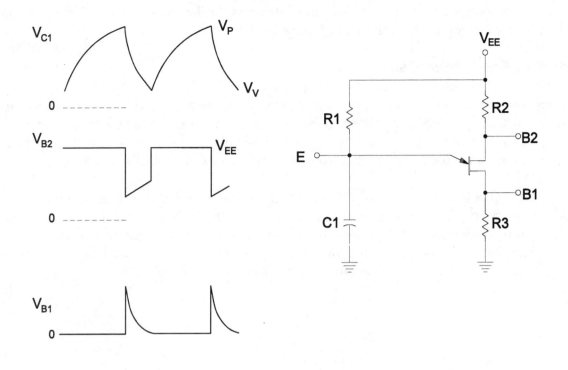

Figure 9-25. Unijunction transistor oscillator and waveforms.

A tunnel-diode monostable multivibrator and its waveforms are shown in *Figure 9-24*. A trigger pulse is required for each output pulse. The output pulse duration is a function of the time constant of the inductance L in the circuit.

A unijunction transistor oscillator and its waveforms are shown in *Figure 9-25*. Its frequency of oscillation is controlled by R1 and C1. When the circuit is turned on, the B1-emitter junction is unbiased; therefore, no current flows through R3. Capacitor C1 charges through resistor R1. As soon as the UJT's threshold voltage is exceeded, it turns on. Capacitor C1 is quickly discharged by the low-impedance PN junction and the UJT quickly turns off. A narrow pulse appears across resistor R3. The emitter waveform is a sawtooth and it can be used for a time delay of up to one minute. The off time for a unijunction transistor is $t_{OFF} = R1C * \ln(1/[1 - n])$, where n is the intrinsic stand-off ratio of the unijunction transistor. A typical value of n is 0.65.

Unijunction transistors only require about 0.4-12 uA of current to turn on. They are therefore useful in high-impedance circuits used to detect small currents. Unijunction transistor oscillators can operate in the 1 Hz to 1 MHz range.

A neon tube oscillator is shown in *Figure 9-26*. A neon tube lights up when the voltage across the electrodes exceeds the ionization potential of the neon gas. When the lamp is not ionized, the neon gas does not glow and it conducts no current.

When power is applied to the neon tube oscillator, capacitor C1 charges. When the capacitor voltage exceeds the threshold voltage of the tube, the neon gas ionizes. The neon tube resistance drops to a very low value. A resistor in series with the neon tube may be needed to prevent the neon tube from being destroyed by the sudden onrush of current. Capacitor C1 discharges until the voltage across the tube drops below the hold or minimum voltage required to ionize the neon gas. As soon as the neon gas is no longer ionized, the tube reverts to its high resistance state, capacitor C1 begins to charge again, and the cycle repeats itself. The voltage across C1 varies between the holding and threshold voltages. The frequency of oscillation is determined by the holding and threshold voltages as well as by components R1 and C1.

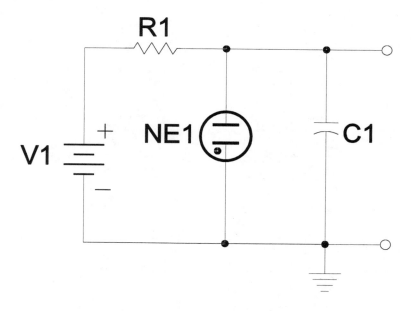

Figure 9-26. Neon tube oscillator.

Problems

Problem 9-1. What characterizes a negative-resistance device?

Problem 9-2. Name some two-terminal negative-resistance devices.

Problem 9-3. Name some three-terminal negative-resistance devices.

Problem 9-4. Name a four-terminal negative-resistance device.

Problem 9-5. Which negative-resistance device is not current controllable? Why?

Problem 9-6. What is the strong point of a tunnel diode?

Problem 9-7. What are the weak points of a tunnel diode?

Problem 9-8. What is the oldest negative-resistance device?

Problem 9-9. What is the strong point of a silicon unilateral switch?

Problem 9-10. What are the weak points of a silicon unilateral switch?

Problem 9-11. Why is the silicon bilateral switch useful in AC circuits?

Problem 9-12. What is the difference between an SCR and a triac?

Problem 9-13. What negative-resistance devices are useful in relaxation oscillator circuits?

Problem 9-14. Compare bistable, monostable and astable modes of operation.

Problem 9-15. What is the difference between positive-feedback and negative-resistance oscillators?

Problem 9-16. What is the off time of a typical unijunction transistor oscillator if R1 = 10 kohm and C1 = 0.001 uF?

Chapter 10

Linear Sawtooth Waveform Generators

There are several linear sawtooth waveform generator circuits. The two most popular are the Miller integrator circuit and the bootstrap circuit, because they are very efficient circuits. The unijunction transistor and the LM555 timer integrated circuit can also be used as linear sawtooth waveform generator circuits.

A sawtooth waveform may be generated by charging and discharging a capacitor, usually by a switching circuit. Recurrent or free-running switching can be provided by a unijunction transistor which is controlled by its emitter voltage. Synchronized recurrent switching is achieved by pulsing the unijunction transistor with a pulse of the required frequency. If a transistor is used to discharge the capacitor then the circuit is operated in a driven mode rather than in a recurrent mode.

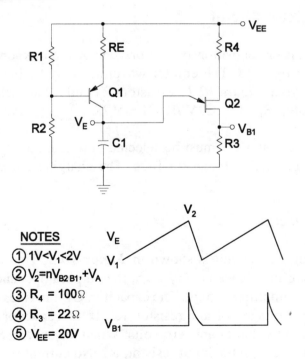

NOTES
1. $1V < V_1 < 2V$
2. $V_2 = nV_{B2\,B1} + V_A$
3. $R_4 = 100\,\Omega$
4. $R_3 = 22\,\Omega$
5. $V_{EE} = 20V$

***Figure 10-1**. Unijunction transistor linear sawtooth generator.*

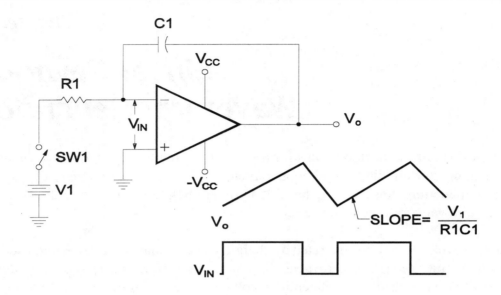

Figure 10-2. *Miller integrator circuit.*

Unijunction Transistor Circuit

The unijunction transistor relaxation oscillator generates an exponential sawtooth waveform as shown in *Figure 9-18*. The emitter waveform may be linearized with a constant current source as shown in *Figure 10-1*. Resistors R1 and R2 are selected to give a current, I_E, in the emitter resistor, R_E: $I_E = C1V_p/R_EC1 = V_p/R_E$.

The base voltage of transistor Q1 must be selected in relation to V_p such that Q1 does not saturate before the unijunction transistor fires. The charging current remains essentially constant.

Miller Integrator Circuit

The basic Miller integrator circuit is shown in *Figure 10-2*. The operational amplifier is ideal; that is, the input impedance is very high, the output impedance is very low, and the gain of the operational amplifier is high. The capacitor, C1, provides feedback and the input current is provided by voltage, V1 and resistor, R1. If the capacitor is initially discharged, then the input and output voltages are zero volts. When the switch, SW1, is closed at t = 0, a current V1/R1 flows through the input resistor, R1. No current flows into the operational amplifier because of its high input impedance. All of the current flows into the capacitor and begins to charge it. The capacitor voltage begins to change at the rate: $dV_o/dt = i/C1 = V1/R1C1$. If the gain of the operational amplifier is very large, V_{IN} remains small regardless of the output voltage. Therefore the current in the capacitor remains a nearly constant $V1/R1$, and the output voltage changes linearly with time.

Bootstrap Sweep Circuit

The bootstrap sweep circuit is similar to the Miller integrator circuit. The bootstrap circuit can generate a sweep voltage that starts at a zero voltage referred to ground. A bootstrap sweep circuit is shown in *Figure 10-3*. The bootstrap circuit gets its name from the fact that the capacitor voltage drop across C1 and the voltage driving resistor R1 rise together. The capacitor voltage drop across C1 raises itself by its "bootstraps".

When transistor Q1 is cut off, it cannot supply base current to transistor Q2. Capacitor C1 is initially discharged. Since Q2 is not conducting, capacitor C1 is charged by the current flow through resistor R1. As capacitor C1 charges, Q2 begins to conduct and the output voltage increases linearly with slope: output slope = Vcc/R1C1. The maximum pulse width of the trigger pulse, $t_p(max)$, is $t_p(max) = Vcc/slope = R1C1$.

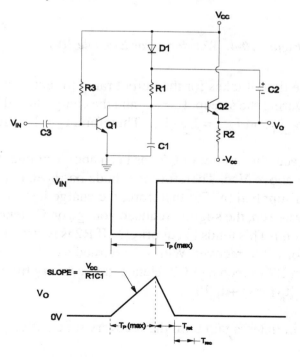

Figure 10-3. Bootstrap sweep circuit.

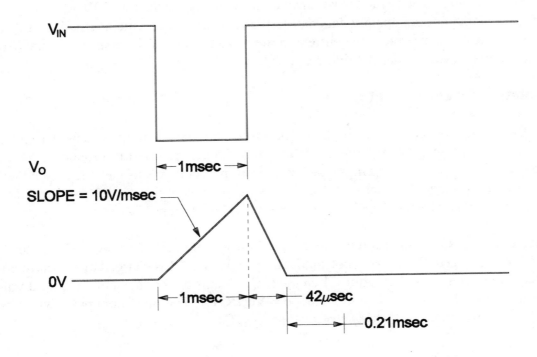

Figure 10-4. Waveforms for Example 10-1.

The retrace time, t_{ret}, is the time it takes for the output ramp to return to zero volts from its maximum value of Vcc. During the sweep, $I_{C1} = I_{R1}$ and the charge stored in C1 is $I_{C1}t_p(max)$. When Q1 turns on, $I_{C2} = h_{FE}I_B$ and $I_{disch} = I_{C2} - I_{R1}$. Therefore, $t_{ret} = I_{R1}t_p(max)/I_{disch}$.

During the sweep and retrace, the emitter of Q2 has been above ground; therefore, the cathode of diode D1 has been above Vcc. Therefore the diode has been cut off and the current flow through R1 has been supplied by C2; therefore, the charge lost by C2 is $I_{C1}[t_p(max) + t_{ret}]$. Immediately after the sweep, the slightly reduced voltage on C2 tends to pull the output emitter slightly above ground. This tends to cut off Q2. If R2 is returned to ground, it would take five time constants for C2 to recover. With R2 returned to -Vcc, we have Vcc/R2 amperes of current flowing in R2 to recharge C2. Hence, the recovery time, t_{rec}, is t_{rec} = charge lost/charging current = $I_{R1}[t_p(max) + t_{ret}]/I_{R2}$.

The sweep should not be initiated again until the recovery time has elapsed.

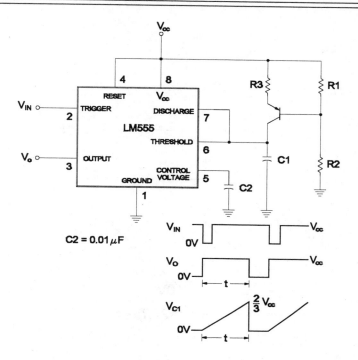

Figure 10-5. *LM555 timer linear sawtooth generator.*

Example 10-1

Analyze the bootstrap sweep circuit of *Figure 10-3* if R1 = 10 kohm, R2 = 2 kohm, R3 = 20 kohm, C1 = 0.1 uF, C2 = 100 uF,

Vcc = 10 volts, -Vcc = -10 volts, and h_{FE} = 50. Draw the waveforms.

t_p(max) = R1C1 = 10,000 x 0.0000001 = 1 mS.

Output slope = Vcc/R1C1 = 10/0.001 = 10 V/msec.

I_{R1} = Vcc/R1 = 10/10,000 = 1 mA.

$I_{C2} = h_{FE}I_B = h_{FE}$Vcc/R3 = 50 x 10/20,000 = 25 mA.

$I_{disch} = I_{C2} - I_{R1}$ = 0.025 - 0.001 = 0.024 = 24 mA.

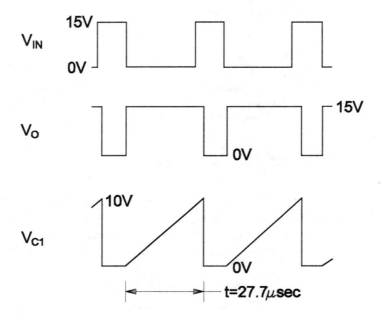

Figure 10-6. *Waveforms for Example 10-2.*

Retrace time: $t_{ret} = I_{R1}t_p(max)/I_{disch} = 0.001 \times 0.001/0.024 = 42$ uS.

Charge lost by C2 is $I_{C1}[t_p(max) + t_{ret}] = 1$ mA(1 mS + 0.042 mS.

$I_{R2} = ABS(-Vcc/R2) = ABS(-10/2000) = 5$ mA.

Recovery time: $t_{rec} = I_{R1}[t_p(max) + t_{ret}]/I_{R2} = 0.001 \times 1.042/0.005 = 0.21$ msec. The waveforms are shown in *Figure 10-4.*

LM555 Timer Linear Sawtooth Circuit

The LM555 timer linear sawtooth generator circuit is shown in *Figure 10-5*. The pull-up resistor of the monostable multivibrator circuit is replaced by a constant current source. The capacitor charges from zero volts to 2/3 Vcc. The linear ramp time interval, t, is t = [2/3VccR3(R1 + R2)C1]/[R1Vcc - V_{BE}(R1 + R2)], where $V_{BE} = 0.6$ volts.

Example 10-2

Draw the waveforms for the LM555 linear sawtooth generator if R1 = 39 kohm, R2 = 100 kohm, R3 = 10 kohm, C1 = 0.01 uF, and Vcc = 15 V. The waveforms are shown in *Figure 10-6*. The linear ramp time interval, t, is:

$$t = [2/3 Vcc R3(R1 + R2)C1]/[R1 Vcc - V_{BE}(R1 + R2)].$$

$$t = 10 \text{ x } 10{,}000 \text{ x } 139{,}000 \text{ x } 0.00000001/585{,}000 - 0.6(139{,}000).$$

$$t = 100 \text{ x } 139 \text{ x } 0.01/585{,}000 - 83{,}400 = 139/501{,}600 = 27.7 \text{ uS}.$$

Problems

Problem 10-1. Why are the Miller integrator and bootstrap circuits the most popular linear sawtooth generator circuits?

Problem 10-2. How is an exponential sawtooth waveform generated?

Problem 10-3. How is a linear sawtooth waveform generated?

Problem 10-4. Calculate the emitter current in a unijunction oscillator where the capacitor is charged by a constant current source. Assume that $V_p = 5$ volts and $R_E = 100$ ohms.

Problem 10-5. Analyze the bootstrap sweep circuit shown in *Figure 10-3* if R1 = 15 kohm, R2 = 10 kohm, R3 = 30 kohm, C1 = 0.2 uF, C2 = 100 uF, Vcc = 15 V and $h_{FE} = 50$. Draw the waveforms.

Problem 10-6. Draw the waveforms for the LM555 linear sawtooth generator if R1 = 47 kohm, R2 = 100 kohm, R3 = 2.7 kohm, C1 = 0.01 uF, and Vcc = 5 V. Assume that $V_{BE} = 0.6$ volts.

Part 2

Oscillator Projects

Chapter 11

Function Generator

A function generator is a waveform generator. It is useful for designing and troubleshooting electronic circuits. This project can generate sine, square and triangle waves at any frequency within the 0.1 Hz to 1 MHz range. The function generator also generates TTL, AM and FM output waveforms.

This function generator is an advanced project. It consists of a dual power supply, a function generator integrated circuit, a high frequency sinewave and triangle wave amplifier, a DC level shifter, a squarewave generator, a TTL output generator and a DC offset and output amplifier.

Circuit Description

The power supply is shown in *Figure 11-1*. Transformer T1 steps down the 117 volt line current to 12.6 volts AC, center-tapped. Each half therefore generates 6.3 volts. The low AC voltage is rectified by the rectifier diode bridge consisting of diodes D1-D4. Capacitors C1 and C2 smooth the DC voltage. Resistor R1 and light-emitting diode D5 are the "power on" indicator components. U1 is a positive five volt regulator and U2 is a negative five volt regulator. Capacitors C3 and C4 improve the transient response of integrated circuits U1 and U2.

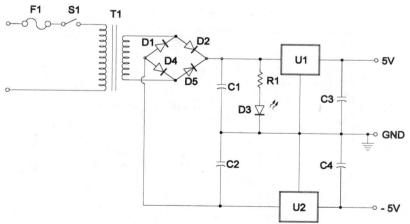

Figure 11-1. Power supply for function generator.

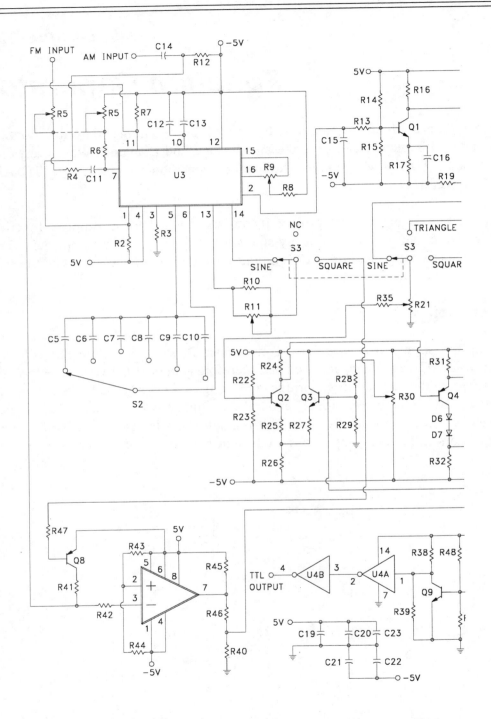

***Figure 11-2**. Schematic of function generator.*

The rest of the function generator circuit is shown in *Figure 11-2*. The brains of the function generator is U3 which is the XR2206 integrated circuit. It consists of an internal voltage-controlled oscillator, sine shaper, multiplier, current switches and a buffer amplifier as shown in *Figure 11-3*. The frequency is a function of the timing resistor connected to pin 7 and the timing capacitor connected to pins 5 and 6. A triangular waveform is generated across the timing capacitor because the timing capacitor charges and discharges through a constant current source set by the timing resistor. When a voltage is applied to the FM input, the control current is varied from pin 7 which in turn varies the output frequency. The squarewave output of the VCO (pin 11) is fed to the collector of transistor Q8. The squarewave is generated internally by the alternate saturation and cut-off conditions of the internal transistor. The triangle wave across the timing capacitor is fed to an internal multiplier and sine shaper circuit. If pins 13 and 14 are open-circuited, the output is a triangle waveform. The output is a sinewave when a resistance is placed between pins 13 and 14. An AM output is obtained when a voltage is applied to the AM input because the gain of the internal amplifier is varied.

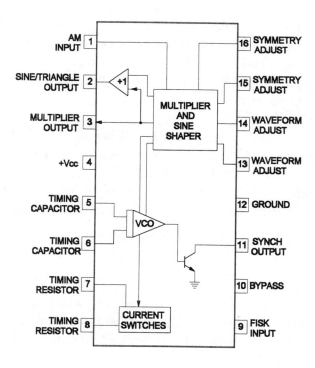

Figure 11-3. Inside the XR 2206 function generator IC.

The output of U3 starts to taper off at 100 kHz. A high frequency amplifier consisting of transistor Q1 and associated components is required for a flat response to 1 MHz. Capacitor C15 attenuates high frequency interference and capacitor C16 increases the high frequency gain of the amplifier because the impedance of C16 decreases as the frequency increases.

The DC level shifter matches the DC level of the sine and triangle waves to that of the squarewave. It also supplies a trigger source for the TTL circuit.

To prevent interference, there is no squarewave generated when switch S3 is in either the sinewave or triangle wave position. Integrated circuit U5 is a voltage comparator. It is required because the output at U3 pin 11 is sensitive to loading and tends to roll-off in amplitude at higher frequencies. When the squarewave swings positive, the voltage at the inverting input exceeds the reference voltage at the non-inverting input causing the comparator output to swing negative. When the squarewave swings negative, the output swings positive because the non-inverting input voltage exceeds the inverting input voltage.

As the DC offset wiper of potentiometer R30 approaches the negative supply, the DC voltage at the base of transistor Q3 is reduced and the DC output voltage is increased. When the DC offset wiper of potentiometer R30 approaches the positive supply, the DC voltage at the base of transistor Q3 is increased and the DC output voltage is decreased. The amplitude of the output waveform is determined by the gain of the amplifier stage consisting of transistor Q4 and its associated components.

The TTL circuit consists of U4 and transistor Q8. When the sine or triangle wave is fed to transistor Q8, it is either saturated or cut-off. A squarewave is generated which is fed to the Schmitt trigger integrated circuit U4. U4 consists of six inverting Schmitt triggers. Only two are required for the function generator. The Schmitt trigger inverters increase the edge speed of its input squarewave slightly. Resistor R35 and potentiometer R9 control the duty cycle of the TTL output.

Construction

The function generator can be built on a piece of perf board. Care must be taken when soldering the connections. All transistors, diodes, and integrated circuits must be properly oriented. It is recommended that sockets for integrated circuits U3, U4 and U5 are used.

The function generator is an advanced project as can be seen from the parts list given in *Table 11-1*. If the reader has limited experience with soldering, he or she should practice soldering on a scrap piece of perf board.

All Resistors Are 1/4W @ 5% Unless Otherwise Noted	All Capacitors Rated At 16 Volts
R1, R4, R6, R8, R35: 1000 ohm	R42, R47: 100k ohm
R2: 68k ohm	R43: 5.6k ohm
R3, R12, R14, R22, R23, R34: 33k ohm	R44, R46: 3.3k ohm
R5: Dual 100k log. potentiometer	C1, C2: 4700 uF
R7, R48: 47k ohm	C3, C4, C13: 0.1 uF
R9: 30k trim potentiometer	C5: 100 uF
R10, R16: 330 ohm	C6: 10 uF tantalum
R11: 1k trim potentiometer	C7: 1 uF tantalum
R13, R15: 12k ohm	C8: 0.1 uF tantalum
R17: 220 ohm	C9: 0.01 uF ceramic
R18: 100 ohm trim potentiometer	C10: 560 pF ceramic
R19: 270 ohm	C11: 47 uF
R20, R41: 4.7k ohm	C12: 10 uF
R21, R30: 5k linear potentiometer	C14: 0.47 uF
R24, R32: 680 ohm	C15: 270 pF ceramic
R25, R27: 56 ohm	C16, C17: 0.002 uF
R26: 1.5k ohm	C18: 0.1 uF
R28: 22k ohm	T1: 12.6 V.C.T.
R29, R36, R39: 10k ohm	D1 - D4: 1N4001
R31, R33: 47 ohm	D5: LED
R37, R38: 2.2k ohm	D6, D7: 1N4001
R40, R45: 470 ohm	U1: LM7805
	U2: LM7905
	U3: XR2206
	U4: 7414
	U5: LM311
	Q1, Q2, Q3, Q5, Q7, Q9: 2N3904
	Q4, Q6, Q8: 2N3906
	F1: 1A fast blow
	S1: SPST
	S2: SP6T
	S3: DP3T
	ICS3: 16-pin IC socket
	ICS4: 14-pin IC socket
	ICS5: 8-pin IC socket

Table 11-1. *Parts list for function generator.*

Capacitors C19-C23 are power supply decoupling capacitors. They should be placed as near as possible to the power supply leads of integrated circuits U3, U4 and U5.

Test and Calibration

If integrated circuits U3, U4 and U5 have been installed, remove them from their sockets. With the power line unplugged, measure the DC resistance from U4 pin 7 to U4 pin 14. If the resistance is greater than 10 ohms, the positive supply is good. If not, check for solder bridges and properly oriented transistors and diodes. Again with the power line unplugged, measure the DC resistance from U4 pin 7 to U3 pin 12. If the resistance is greater than 10 ohms, the negative supply is good. If not, check for solder bridges and properly oriented transistors and diodes. With the power line plugged into household current, measure the DC voltage drop across U4 pin 7 to U4 pin 14. It should measure five volts. Measure the DC voltage drop across U4 pin 7 to U3 pin 12. It should measure minus five volts. When all is correct, unplug the line cord and insert U3, U4 and U5 into their sockets. Obviously, if the power supplies are not functioning, immediately unplug the function generator until the problem is found and corrected. Check for solder bridges and properly oriented diodes, transistors and integrated circuits U1 and U2.

Rotate potentiometers R5, R9, R11, R18 and R30 to mid-position and potentiometer R21 fully clockwise. Set S3 to the sinewave position and set S2 to the "1k" position. With an oscilloscope connected between the output and ground terminals, verify that the function generator is generating an approximate sinewave of about 1 kHz. If there is no output, find and correct the problem before calibrating the function generator.

Rotate the function switch S3 to the squarewave position. Adjust the DC offset potentiometer R30 until the squarewave is centered relative to ground. With R30 set to the sinewave position, adjust R18 until the sinewave is also centered relative to ground. If the sinewave is distorted, adjust R11 for minimum distortion with R9 set at midpoint. Then adjust R9 for minimum distortion without disturbing the R11 setting. Verify that the sinewave is still centered relative to ground. If not, readjust potentiometer R18. The function generator should now be functioning properly and it is ready for years of trouble free use.

Using the Function Generator

The sine, triangle or squarewave is available at the output terminal. The function switch is set to the appropriate position and the amplitude control R21 is set for the required peak-to-peak waveform voltage. The frequency control R5 and the range switch S2 are set for the required output waveform frequency.

The TTL output is always available at the TTL output terminal. The duty cycle can be adjusted by the symmetry adjust control R9. The TTL output waveform has a rise time of approximately 50 nS and a fall time of about 20 nS.

An AM waveform is available at the output terminal when a modulating signal of at least 30 Hz is fed to the AM input terminal. The function switch must not be on the squarewave setting because a squarewave cannot be modulated.

An FM waveform is available at the output terminal when a modulating signal of at least 10 Hz is applied to the FM input terminal. The function switch must not be on the squarewave setting because a squarewave cannot be modulated. Capacitor C11 will be damaged if the modulating signal has a DC offset greater than 13 volts or less than -2 volts relative to ground.

Chapter 12

Capacitance Meter

This capacitance meter autoranges from 1 pF to 1 uF and from 1 uF to 4000 uF, and it updates readings automatically. The capacitance meter is useful for determining the values of unmarked capacitors. The meter's accuracy is about one percent.

Capacitors are usually measured on an AC bridge by balancing the reactance of known components against the reactance of an unknown capacitor at a fixed frequency. This meter measures capacitance by measuring time.

The unknown capacitor is first charged to a known voltage. The capacitor is then discharged to another known voltage through a fixed resistance. The discharge time is directly proportional to the unknown capacitance. The discharge time is determined by this capacitance meter.

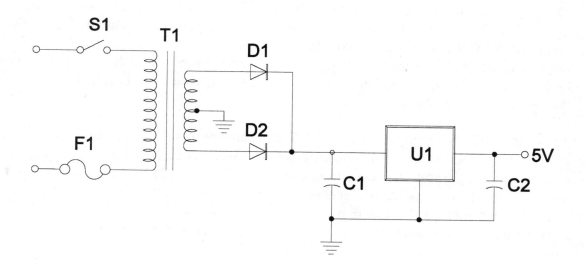

Figure 12-1. Power supply for the capacitance meter.

Circuit Description

The schematic of the power supply of the capacitance meter is shown in *Figure 12-1*. Transformer T1 steps the household line voltage down to 12.6 volts center-tapped. Diodes D1 and D2 rectify the low AC voltage to a pulsating DC voltage. The positive regulator integrated circuit U1 provides a constant five volts to the rest of the capacitance meter circuitry. Capacitor C1 smooths out the pulsating DC voltage and C2 improves the transient response of U1. The rest of the capacitance meter is shown in *Figure 12-2*.

The unknown capacitance becomes the timing capacitor for U2 which is configured as an astable multivibrator. When switch S2 is in the nF position, components R1, R2 and Cx determine the discharge time of Cx. The discharge time of Cx is determined by components R3, R4 and Cx when S2 is in the uF position.

A second LM555 timer integrated circuit, U6, is also configured as an astable multivibrator. It is part of an autocycling circuit which automatically updates the capacitance measurement. The autocycling circuit is supplied with a 1.4 MHz reference clock. The clock is generated by integrated circuit U4 which is configured as a Colpitts oscillator. Components L1 and C7-C10 are the tank circuit for the Colpitts oscillator.

The outputs of the astable multivibrators and the Colpitts oscillator are combined into U5 and U8, which are dual D flip-flops. One-half of U5 synchronizes the output of U2 with the output of the Colpitts oscillator providing dual-phase outputs. The other half of U5 and integrated circuit U8 select one discharge pulse U2 when the output of the autocycler astable multivibrator U6 is high. Astable multivibrator U6 is disabled by the flip-flops until the discharge pulse is completed.

The Colpitts oscillator output is gated by U7 and then it passes to the counting stages during one discharge period of the unknown capacitor each measuring interval. Monostable multivibrator U3 resets the decade counters, U15, U18-U20 and dividers U11-U13. Monostable multivibrator U3 is triggered by the leading edge of the synchronized discharge pulse. When switch S2 is in the nF position, the reset pulse width of U3 is controlled by the zero trimmer potentiometer R10. This eliminates the effect of stray capacitance on the measurement.

The gated Colpitts oscillator output is divided by decade counters U11-U13. The counter outputs are fed to the tri-state logic switch U9 which passes the pulse train to decade counter U20. Overflow pulses generated by counter U20 are passed on to decade counters U19 and

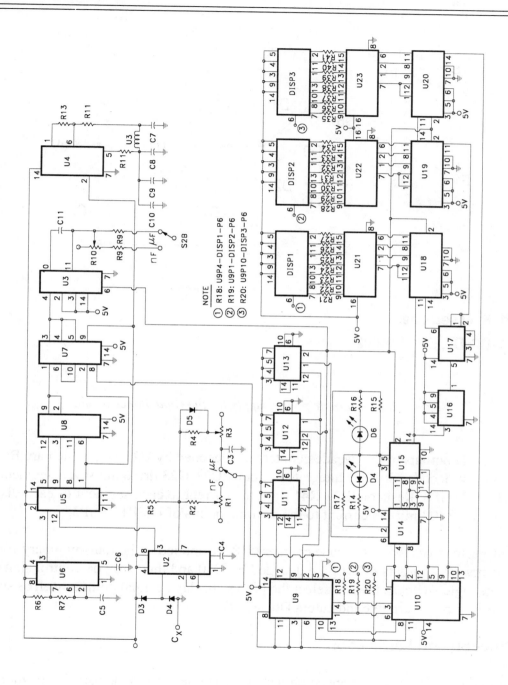

Figure 12-2. Schematic of capacitance meter.

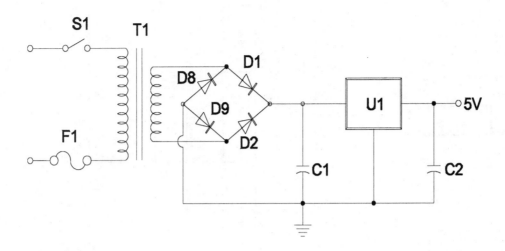

T1: 6.3V secondary

D1, D2, D8, D9: 1N4002

Figure 12-3. Alternate power supply for the capacitance meter.

U18. The BCD outputs of the decade counters are decoded by U21-U23 which are BCD to seven-segment decoder/drivers. Integrated circuits U21-U23 drive the seven-segment displays DISP1-DISP3. Resistors R21-R41 limit the current to each segment of each display. Resistors R18-R20 limit the current to the decimal points of DISP1-DISP3.

Overflow pulses from the last decade counter U18 are applied to the four-bit binary counter U15. The inverted outputs of U14 are decoded by U10 and thus provide control signals to the tri-state logic switches U9 and select the proper display decimal point as well as sink or block current from overrange indicators D6 and D7.

Construction and Verification

The capacitance meter is an advanced project, as can be seen from the parts list of *Table 12-1*. The meter can be built on a piece of perf board. This project can also be wire-wrapped.

The capacitance meter can be completely built and then tested. It is recommended that this project is built and tested in stages.

The power supply should be the first section that is built and tested. The power supply circuit is shown in *Figure 12-1*. If the reader is using a transformer without a center tap, diodes D1 and D2 should be replaced by the diode rectifier bridge consisting of diodes D1, D2, D8 and D9 as shown in *Figure 12-3*. Once the five volt power supply is built and working, the circuits of *Figure 12-2* can be built and verified.

The astable multivibrator circuit of U2 is the first section to be built. A squarewave should be present at U2 pin 3 when a capacitor is connected to the Cx terminals. Verify that this stage works for both settings of switch S2.

The oscillator, synchronizer and reset circuits of U3-U8 should be built next. Do not install capacitors C9 and C10 yet. A squarewave should be present at U6 pin 3 and at U4 pin 2. There should be a squarewave at U7 pin 8 and at U3 pin 6. There is also a squarewave at U5 pin 12 as long as Cx is connected to astable multivibrator U2.

The autoranging circuit is built next. It consists of integrated circuits U9-U20 and associated components. There should be squarewaves or pulses at U11 pins 1 and 12, at U12 pins 1 and 12 as well as at U13 pins 1 and 12. The overflow light-emitting diodes should illuminate only in the overflow condition.

The display circuits should be built last and it consists of BCD to seven-segment decoder/ drivers U21-U23, DISP1-DISP3 and associated components. When there is no capacitor connected to the Cx terminals, the display should indicate a "000" readout within a couple of seconds, if potentiometers R1, R3 and R10 are rotated counterclockwise and S2 is in the nF position.

Calibration and Use

When the ZERO potentiometer R10 is rotated clockwise, the display should indicate a few picofarads. Rotate R10 until the display indicates "000". Connect a capacitor known to be 0.68 uF to the meter. Set switch S2 to the uF position and adjust potentiometer R3 until the display indicates "0.68". Set S2 to the nF position and adjust R1 until the display indicates "680". If the required display is not obtainable, install capacitors C9 and/or C10 and repeat the calibration procedure.

All Resistors Are 1/4W At 5% Unless Otherwise Noted	All Capacitors Rated At 16 Volts
R1: 100k trimmer potentiometer	C1: 4700 uF
R2: iM ohm @ 1%	C2: 0.1 uF
R3: 100 ohm trimmer potentionmeter	C3: 0.0033 uF
R4: 1k ohm @ 1%	C4, C6, C7: 0.01 uF
R5: 1k ohm	C5: 4.7 uF
R6, R7: 100k ohm	C8: 820 pF
R8, R9: 1.5k ohm	C9: 470 pF
R10: 25k ohm linear potentiometer	C10: 220 pF
R11 - R13: 100 ohm	C11: 0.005 uF
R14, R15: 3.3k ohm	U1: LM7805
R16 - R41: 470 ohm	U2, U6: LM555 timer
D1 - D2: 1N4002	U3, U16: 74121
D3 - D5: 1N4154	U4, U14: 7404
L1: 12 uH	U5, U8: 7474
S1: SPST	U7, U10, U17: 7400
S2: DPDT	U9: 74125
F1: 1/4A fast blow	U11 - U13, U18 - U20: 7490
T1: 12.6V C.T. secondary	U15: 7493
DISP1 - DISP3: DI707 common-anode	U21 - U23: 7447
D6 - D7: LED	

Table 12-1. Parts list for the capacitance meter.

Each time the capacitance meter is used with S2 in the nF position, the display should be "zeroed" with the ZERO potentiometer. When S2 is in the uF position, no zeroing is required. The unknown capacitor is connected to the Cx terminals with the positive capacitor lead connected to the positive Cx terminal. The unknown capacitor must be discharged to protect the input circuitry of the capacitance meter.

If the unknown capacitance is more than 1000 nF, place S2 in the uF position. If the unknown capacitance is less than 1000 nF, place S2 in the nF position. If the unknown capacitance is greater than 1000 uF, place S2 in the uF position. The capacitance can be determined by observing the overrange LEDs. If only the upper LED glows, the unknown capacitance is 1000 uF. If only the lower LED glows, the unknown capacitance is 2000 uF. If both LEDs glow, the unknown capacitance is 3000 uF. If the sequence repeats, then the unknown capacitance is 4000 uF or greater, depending upon how often and what part of the sequence repeats.

Capacitors with high leakage will not charge to the reference voltage and therefore will not trigger the discharge cycle. When S2 is in the nF position and the extreme left decimal point is lit, treat the display as reading the unknown capacitance in pFs. A display of ".050" should be read as 50 pF.

Chapter 13

Frequency Counter

This frequency counter can measure frequencies from 1 Hz to about 30 MHz. An input attenuator is not necessary because the input circuit can accept a wide range of input amplitudes. A frequency counter is useful for designing and testing electronic circuits.

The signal being measured must be at least 100 mV in amplitude and must not exceed 150 volts in amplitude at 100 kHz, 100 volts in amplitude at 1 MHz, or 50 volts in amplitude at 10 MHz. The time base frequency of the frequency counter is 1 kHz. There are only two switches to operate, the power switch and the kHz/MHz range switch. The input impedance of the frequency counter is 1 megohm shunted by a capacitance of less than 20 pF. The counter therefore does not load the circuit under test.

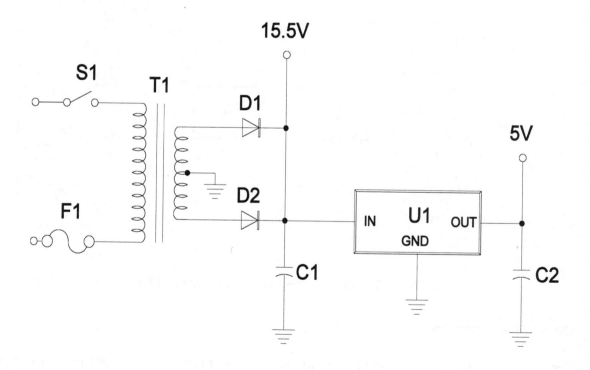

Figure 13-1. *Power supply for the frequency counter.*

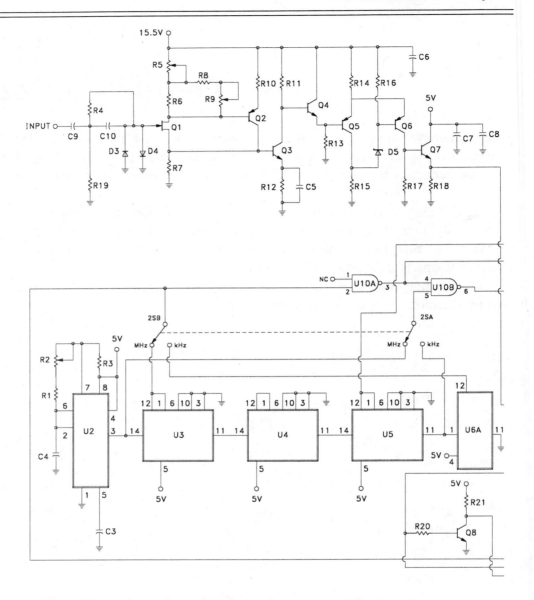

Figure 13-2a. Schematic of the frequency counter. (Continues)

Circuit Description

The power supply for the frequency counter is shown in *Figure 13-1*. Transformer T1 steps down the line voltage to 22 volts center-tapped. This low AC voltage is rectified by diodes D1 and D2 and smoothed by capacitor C1. The unregulated voltage powers the input circuit

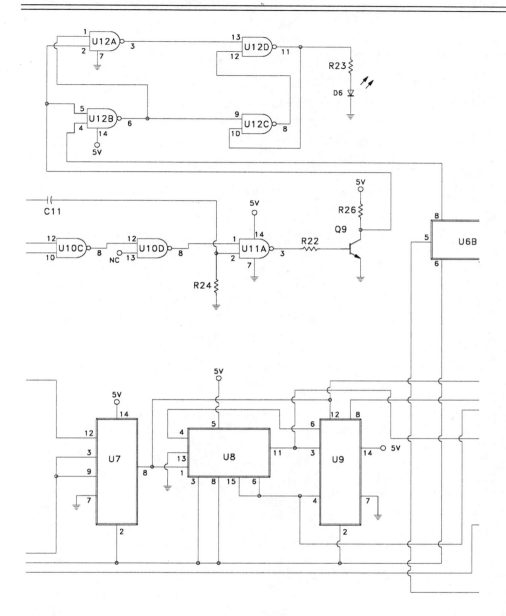

Figure 13-2b. *Schematic of the frequency counter. (Continues)*

which consists of transistors Q1-Q7 and associated components. The regulated five volt supply powers the rest of the frequency counter as shown in *Figure 13-2*. Integrated circuit U1 is the five volt regulator and capacitor C2 improves the transient response of voltage regulator U1.

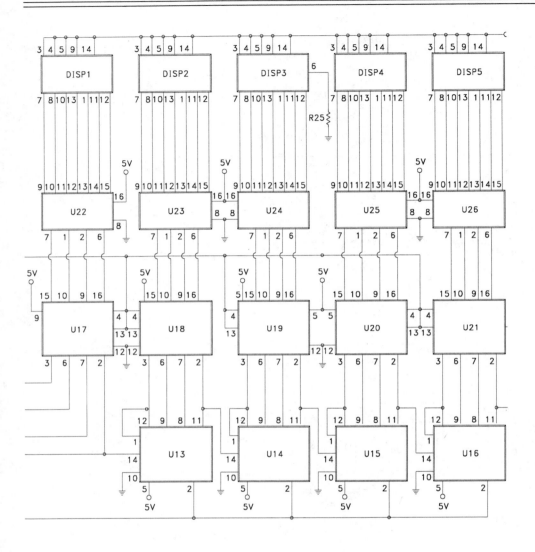

Figure 13-2c. Schematic of the frequency counter. (Concluded)

The input amplifier and Schmitt trigger circuits consist of transistors Q1-Q7 and associated components. They shape the input into a squarewave and feed it to decade counter U7. Capacitor C9 removes any DC voltage from the input signal. C10 prevents attenuation of high frequency signals. The input buffer Q1 is protected from input spikes by resistor R4 and diodes D3 and D4. Transistors Q1 and Q2 are direct coupled with 100% negative feedback to provide unity gain, wide bandwidth, high input impedance and low output impedance. Transistors Q3 and Q4 provide gain and isolation between the input circuit and the Schmitt trigger.

The Schmitt trigger consists of Q5 and Q6. It produces a squarewave each time it is triggered and reset. The threshold of the Schmitt trigger is adjusted by potentiometers R5 and R9. Transistor Q7 serves as a buffer to isolate the Schmitt trigger and the TTL (transistor-transistor logic) counter circuit.

Astable multivibrator U2 generates a 1 kHz squarewave. A scaler which consists of integrated circuits U3-U5 generates a 1 Hz squarewave. The range switch S2 selects either the 1 Hz or the 1 kHz squarewave as the time base. The time base is used for the gating, reset and memory circuit. It is also used to transfer the count to the BCD to seven-segment decoder/drivers U22-U26 and the seven-segment displays DISP1-DISP5.

The positive-going reset pulse is generated at the collector of transistor Q8. The memory transfer pulse is generated at the collector of transistor Q9.

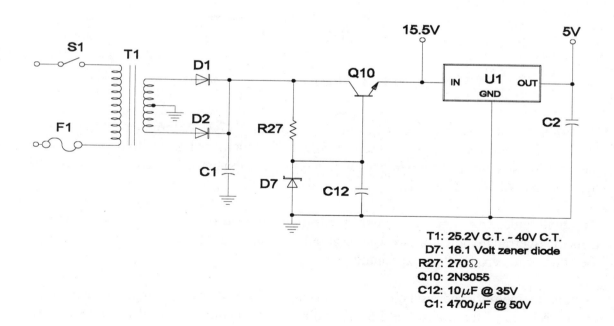

Figure 13-3. Alternate power supply for the frequency counter.

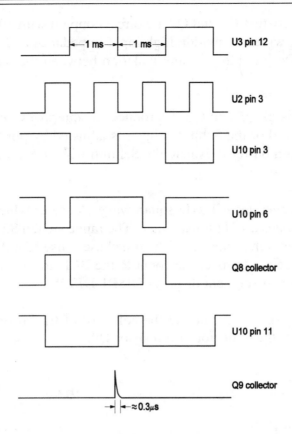

Figure 13-4. MHz range timing diagram.

When the transfer pulse is applied to memory latches U17-U21, they accept the accumulated BCD count from decade counters U13-U16. The memory latches hold this count until the next transfer pulse. The memory latch outputs are connected to the BCD to seven-segment decoder/drivers U22-U26.

Any tenth pulse from decade counter U16 triggers the overrange detector U12 which controls the overrange indicator D6. The reset pulse occurs after the transfer pulse and it resets the counter to zero for the next counting cycle.

The gating, memory and reset circuit controls the times that the input signal is gated into the counting circuits, the times that the accumulated information is passed from the counting circuits to the display circuits and the times that the counter circuits are reset to zero to start a new counting cycle.

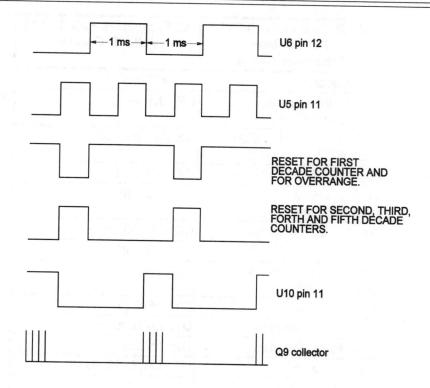

Figure 13-5. Timing diagram for the kHz range.

The first decade counter consists of integrated circuits U7-U9. It is configured as an asynchronous BCD counter. The flip-flops are triggered by negative-going pulses. Flip-flop U7 is toggled by the signal from the Schmitt trigger circuit which consists of transistors Q5 and Q6.

Integrated circuits U13-U16 are configured as the second to fifth decade counters, respectively. These counters are also asynchronous BCD counters. Unlike the first decade counter, the second to fifth decade counters require no gating because the first decade counter must be counting and producing a carry pulse.

Construction and Verification

The frequency counter may be built on a piece of perf board. It is recommended that the project is wirewrapped and that it is built and tested in sections.

All Resistors Are 1/2W @ 5% Unless Otherwise Noted	All Capacitors Rated At 25 Volts or Greater
R1: 12k ohm	C1: 4700 uF
R2: 5k trimmer potentiometer	C2: 0.1 uF
R3, R11, R13, R20, R21, R22, R26: 1k ohm	C3, C7: 0.01 uF
R4: 100k ohm	C4: 0.047 uF
R5: 3k trimmer potentiometer	C5: 0.68 uF
R6, R7: 220 ohm	C6, C8: 10 uF
R8: 150 ohm	C9: 0.47 uF
R9: 500 ohm trimmer potentiometer	C10: 10 pF
R10: 68 ohm	C11: 680 pF
R12, R18: 100 ohm	U1: LM7805
R14, R24: 390 ohm	U2: LM555
R15: 47 ohm	U3 - U5, U-13 - U16: 7490
R16: 1800 ohm	U6: 7473
R17: 270 ohm	U7: 74H102
R19: 1M ohm	U8: 7476
R23, R25: 470 ohm	U9: 7472
D1, D2: 1N4002	U10 - U12: 7400
D3, D4: 1N4149	U17 - U21: 7475
D5: 5.6 volt zener diode	U22 - U26: 7447
D6: LED	DISP1 - DISP5: DL707
Q1: SFC2912	T1: 22VCT, 1A secondary
Q2: 2N4121	S1: SPST
Q3, Q7: 2N3563	S2: DPDT
Q4: MPS6520	F1: 1/4A slow blow fuse
Q5, Q6: 2N4258A	AC1: AC line cord
Q8: 2N3393	MISC: Cabinet, wire, etc.
Q9: 2N2369	

Table 13-1. Parts list for the frequency counter.

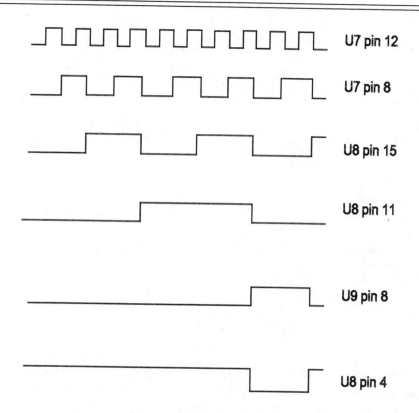

Figure 13-6. First decade counter timing diagram.

The power supply may be built on a separate piece of perf board. Diodes D1 and D2 and capacitor C1 must be properly oriented. Integrated circuit U1 must be properly connected into the circuit. It is recommended that a TO-3 case LM7805 regulator is used. Verify that the voltages are correct. If the required power transformer T1 is not available, a transformer with a 25.2 volt to 40 volt center-tapped secondary may be used with a series-pass transistor regulator as shown in *Figure 13-3*.

The astable multivibrator consisting of U2 and associated components should be built next. There should be a 1 kHz squarewave present at U2 pin 3. The circuits consisting of U3-U12 and transistor switches Q8 and Q9 should be built next. There should be a 100 Hz squarewave at U3 pin 11, a 10 Hz squarewave at U4 pin 11, and a 1 Hz squarewave at U5 pin 11.

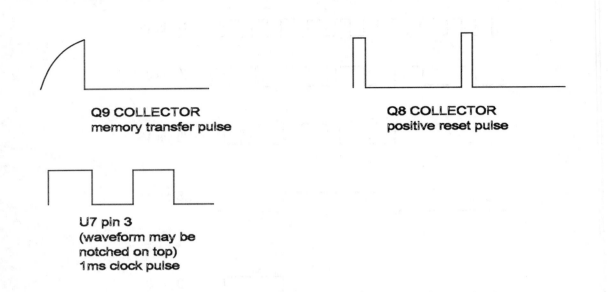

Figure 13-7. *Other important circuit waveforms.*

The input circuit consisting of transistors Q1-Q7 and associated components should be built next. When an input signal is present, there should be a squarewave of the same frequency as the input signal present at the base and at the emitter of transistor Q7.

Finally, the circuits consisting of integrated circuits U13-U26 and displays DISP1-DISP5 should be built. The completed frequency counter can be tested for proper operation. The timing diagram for the MHz range is shown in *Figure 13-4* and the timing diagram for the kHz range is shown in *Figure 13-5*. The timing diagram for the first decade counter is shown in *Figure 13-6*. *Figure 13-7* shows the memory transfer pulse, the positive-going reset pulse and the 1 millisecond clock pulse.

If any problems are encountered, verify that all transistors and diodes are properly oriented. Five volts should be present at the correct pins of all of the TTL logic integrated circuits and displays DISP1-DISP5. If the project was not wirewrapped, check for solder bridges and cold solder joints.

Calibration and Use

The time base oscillator is tuned with no input signal applied to the frequency counter. Connect an oscilloscope between U2 pin 3 and ground. Adjust trimmer potentiometer R2

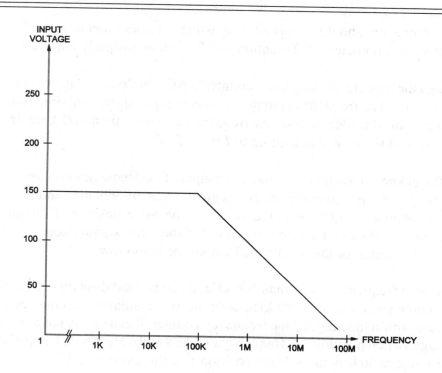

Figure 13-8. Maximum input voltage derating curve.

until the clock pulse has a period of exactly 1 millisecond. If there is no oscilloscope available, connect a multimeter between U5 pin 11 and ground. Set the multimeter to its 10 volt DC range. The time base frequency at U5 pin 11 is 1 Hertz. It switches from 0 volts to approximately 5 volts, sixty times per minute. It also switches from approximately 5 volts to zero volts, sixty times per minute. Adjust trimmer potentiometer R2 until there are exactly sixty 0 volt to 5 volt transitions each minute, or, until there are exactly sixty 5 volt to 0 volt transitions each minute. The time base clock is now properly adjusted.

An input signal of known frequency and an amplitude of about 500 millivolts is connected to the input of the frequency counter. Adjust trimmer potentiometer R5 until the frequency counter displays the frequency of the input signal. Reduce the amplitude of the input signal and again adjust R5 until the frequency counter displays the frequency of the input signal. Repeat this procedure until the input signal amplitude is too low to obtain a correct display on the frequency counter. Potentiometer R5 is now properly adjusted.

The above procedure should be repeated adjusting trimmer potentiometer R9. Do not readjust trimmer potentiometer R5. Potentiometer R9 is now properly adjusted.

Even though the input to the frequency counter is AC coupled, the DC input level should not exceed 200 volts. The frequency counter can accept input signal amplitudes of 150 volts AC at frequencies of 100 kHz or less. At frequencies greater than 100 kHz, the input signal amplitude should be derated according to *Figure 13-8*.

Connect the unknown frequency signal to the input of the frequency counter. If the overrange LED lights up, the input signal frequency is greater than 99.999 kHz and the range switch is in the kHz position. Switch the range switch to the MHz position. If the display changes constantly in a random manner, the frequency of the input signal exceeds the capability of the frequency counter, or the input signal amplitude is too low.

If the unknown frequency is less than 100 kHz, it can be read directly to a resolution of 1 Hz when the range switch is set in its kHz position. If the unknown frequency is greater than 100 kHz and within the range of the frequency counter, it can be read to a resolution of 1 Hz in both range switch positions. A frequency of 24,589,316 Hz would be displayed as 24.589 when the range switch is in its MHz position and the overrange LED is off. If the range switch is in its kHz position, the display would be 89.316 and the overrange LED is on.

Chapter 14

Solar Powered Generator

The solar powered generator is powered by solar cells and generates household line current. The power capability of this project is determined by the transformer and power transistors. The secondary of the transformer also determines the solar cell voltage.

The solar powered generator can be used to operate television sets, VCRs and other household appliances. The project can be used to power emergency equipment during daylight power failures. If the solar cells are also used to charge a backup battery, the project can be used to power emergency equipment during nighttime power failures.

This project is easy to build and should provide years of trouble-free service.

Circuit Description

The solar generator is a power astable multivibrator as shown in *Figure 14-1*. The astable multivibrator consists of transistors Q1 and Q2 and timing components R3, R4, C1 and C2. The duty cycle of the astable multivibrator must be 50%, just like household current. Therefore, components R3 = R4 and C1 = C2. These timing components are selected for an operating frequency of about 70 Hz. The 70 Hz frequency, rather than 60 Hz, is used to prevent saturation of the transformer core.

Residential AC outlets have a ground side and a "hot" side. The hot side alternates from -170V AC to +170V AC. The solar generator output is similar. During one-half of the AC cycle, one side is near ground potential while the other side is at +170V AC. The situation is reversed during the other half of the AC cycle.

The power transistors Q3 and Q4 are configured as inverter switches. Resistors R5 and R6 limit the base drive current to the power transistors. The power switches drive transformer T1 which is a power transformer used as a step-up transformer. The power rating of the transformer must not exceed the power dissipation capabilities of the power transistors. The secondary winding of the power transformer must be center-tapped as shown in the schematic of *Figure 14-1*.

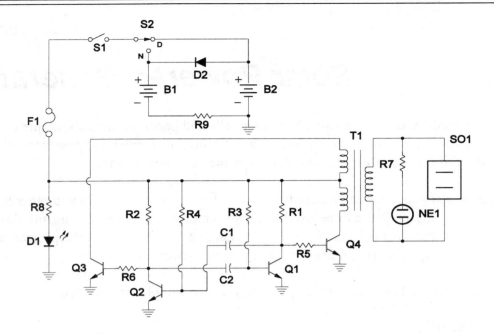

Figure 14-1. Schematic of the solar powered generator.

The 2N3055 power transistor can dissipate 115 watts of power when mounted on a suitable heat sink. It can also pass 15 amperes of collector current when properly heat sunk to prevent the junction temperature of the power transistors from exceeding 200 degrees Celsius. At 12 volts the 2N3055 can handle about 9 amperes of current. However, at this extreme, huge heat sinks are required. Select a power transformer with a secondary current output of 5 amperes or less.

The secondary voltage of the transformer determines the required solar cell and optional backup battery voltages. The selected transformer should have as low a secondary voltage as possible to reduce the required solar cell and optional backup battery voltages.

The backup battery B1 can be charged during the day by solar cell(s) B2, when switch S1 is in the open position. Switch S2 can be in either position. The project is powered during the day by B2 when switch S2 is in the "D" position and S1 is in the closed position. The project is powered during the night by backup battery B1 when S2 is in the "N" position and switch S1 is in the closed position. Diode D2 prevents current from entering the solar cell(s) at night. Resistor R9 limits the charging current for the backup battery B1 and R9 is selected to suit the requirements of the backup battery used in this project.

Solar cells can be connected in series for a higher voltage and in parallel for a higher current. For example, if a solar cell is rated at 0.5 volts and 0.2 amperes, two cells may be connected in series for an output of 1 volt. Two solar cells may be connected in parallel for an output of 0.4 ampere. Four cells may be connected in series/parallel for an output of 1 volt and an output of 0.4 ampere. Several solar cells may be connected in series/parallel for any required combination of output voltage and current.

A DC power-on indicator consists of resistor R8 and light-emitting diode D1. An AC power-on indicator consists of resistor R7 and neon bulb NE1. Switch S1 is used to turn the project on or off. Fuse F1 is a safety device used to protect the solar cell(s) and the optional backup battery in the unlikely event that the solar power generator fails.

Construction

The solar powered generator may be built on a piece of perf board. The parts list for the project is given in *Table 14-1*. Power transistors Q3 and Q4 must be adequately heat sinked in order to meet the secondary current requirement of the power transformer. If the solar power generator is housed in a metal cabinet, the cabinet may be used as a heat sink for the power transistors. Use mica insulators to electrically isolate the power transistors from the metal cabinet.

The backup battery and the solar cell(s) voltages should equal the power transformer secondary voltage rating. For example, if the power transformer secondary is rated at 12.6 volts center-tapped, then B1 and B2 should be 12.6 voltage sources.

Parts placement is not critical. Keep all leads as short as possible. Make sure that the transistors are properly oriented and that there are no solder bridges or cold solder joints.

Testing

Connect the solar power generator to a suitable power source. Observe the waveforms at the collectors of Q1 and Q2 on an oscilloscope. A 70 Hz squarewave with a duty cycle of 50% should appear on the oscilloscope. The two waveforms should be identical but out of phase by 180 degrees. If not, verify the wiring of Q1, Q2, R1-R4, C1 and C2. When everything checks out, plug an AC powered radio into the solar generator. Turn on the radio and the solar powered generator. The radio should work. If not, check the wiring of Q3, Q4, R5, R6 and T1. The AC and DC power-on indicators should light up as well. The solar powered generator is now fully functional and it should give years of trouble-free service.

All Resistors Are 1/4W At 5% Unless Otherwise Noted
R1, R2, R5, R6, R8: 1 kohm
R3, R4: 22 kohm
R7: 100 kohm
R9: See text.
C1, C2: 0.47 uF
Q1, Q2: 2N2222A
Q3, Q4: 2N3055
D1: LED
D2: 1N5404
NE1: Neon bulb for 117 volt applications.
T1: Power transformer, see text.
F1: 5A slow blow.
S1: SPST
S2: SPDT
B1: Back-up battery, see text.
B2: Solar cells, see text.
SO1: Chassis mount AC socket.

Table 14-1. Parts list for the solar powered generator.

Chapter 15

Electronic Siren

The electronic siren produces a wailing sound. This project can be used in many applications where a warning sound is required. The siren can be powered by a nine-volt battery, a car battery, or by a nine-volt to twelve-volt AC operated power supply.

The electronic siren is easy to build and use. It can be housed in any suitable enclosure. The small size of the project allows it to be hidden from view, a useful feature for alarm systems.

Circuit Description

The electronic siren produces a high pitched sound which gradually decreases until power is reapplied to the siren. Power is applied at regular intervals by U1 which is configured as an astable multivibrator and by U2, which is a quad bilateral electronic switch. The schematic is shown in *Figure 15-1*.

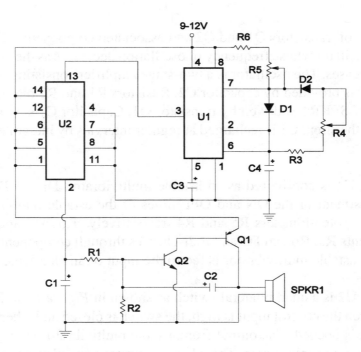

Figure 15-1. Schematic of the electronic siren.

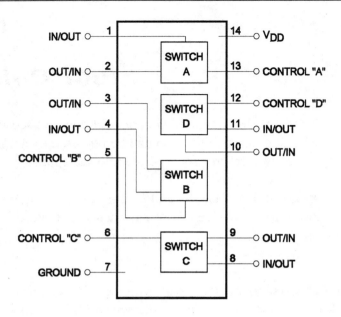

Figure 15-2. Inside the CD4066 quad bilateral switch.

The siren consists of transistors Q1 and Q2 and associated components. The siren is a voltage-controlled oscillator whose frequency of oscillation decreases as the voltage drop across capacitor C1 decreases. The oscillator is a two-stage amplifier consisting of Q1 and Q2. The positive feedback is provided by capacitor C2. Resistors R1 and R2 bias transistor Q2. The eight ohm speaker SPKR1 is driven by transistor Q1. Capacitor C1 stores voltage and discharges to power the siren. C1 is recharged at regular intervals by integrated circuits U1 and U2.

Integrated circuit U1 is configured as an astable multivibrator. Diodes D1 and D2 permit independent adjustment of the ON and OFF times of the astable multivibrator output by adjusting trimmer potentiometers R5 and R4, respectively. Timing capacitor C4 charges through components R5, R6 and D1 and it discharges through components R3, R4 and D2. The output of the astable multivibrator is fed to the input of an electronic CMOS switch.

Integrated circuit U2 is a quad bilateral switch as shown in *Figure 15-2*. Each switch has a control input. When the control input is high, the switch is closed and when the control input is low, the switch is opened. The output from astable multivibrator U1 is used to pulse the control input of an electronic switch. This alternately opens and closes the electronic switch which in turn charges capacitor C1 at regular intervals.

All Resistors Are 1/4W At 5% Unless Otherwise Noted
R1: 75k
R2: 62k
R3: 1M
R4: 470k trimmer potentiometer
R5: 100k trimmer potentiometer
R6: 47k
C1: 47 uF at 16 volts
C2: 0.022 uF
C3: 0.01 uF
C4: 4.7 uF tantalum
U1: LM555
U2: CD4066
Q1: GE-3 or 2N1039 or equivalent
Q2: GE-10 or MPSA05 or equivalent
D1, D2: 1N4148
SPKR1: 8 ohm speaker

Table 15-1. Parts list for the electronic siren.

The quad bilateral switch U2 is a CMOS integrated circuit. All unused inputs of a CMOS device must be tied to either the power supply or to ground.

Construction, Verification and Use

The electronic siren can be built on a piece of perf board. Care should be taken to avoid solder bridges and cold solder joints. All transistors, diodes, integrated circuits and capacitors must be properly oriented. The parts list for the siren is given in *Table 15-1*.

A squarewave should be present at U1 pin 3 and U2 pin 13. If not, check the orientation of diodes D1 and D2. Power should be available at U2 pin 2 whenever the square wave at U1 pin 3 and U2 pin 13 is high. If not, check the wiring of integrated circuit U2. There should be a wailing sound emitted from the speaker. If not, check the orientation of transistors Q1 and Q2 as well as capacitor C1.

When the electronic siren is working properly, trimmer potentiometers R4 and R5 may be adjusted to vary the sound effects to suit the user's needs.

The GE-10 transistor is substituted by an MPSA05 transistor which is available at most electronic parts stores. The GE-3 transistor or equivalent may be harder to obtain. Some GE-3 equivalents include 2N141, 2N143, 2N1039, 2N1040, 2N1043, 2N1044, 2N1437, 2N1438, 2N1611, 2N1612, 2N2071, 2N2072, 2N2565, 2N2566, 2N2567, and HEP238.

Chapter 16

Two Electronic Organs

A monophonic organ is an organ where only one note at a time may be played. The first electronic organ project is a two-octave monophonic organ.

A polyphonic organ is an organ where several notes may be played simultaneously. Chords can be played on a polyphonic organ. The second electronic organ project is a two-octave polyphonic organ.

Both organs cover the musical range from C (262 Hz) to C (1047 Hz), with twelve notes per octave. It therefore includes sharps and flats. *Table 16-1* lists the notes, frequencies and time periods for each of the twenty-five musical notes that are available on both electronic organs. Both electronic organs feature tremolo and pitch control circuits.

Circuit Description

The power supply used for both electronic organs is shown in *Figure 16-1*. Transformer T1 is a center-tapped step-down transformer. Diodes D1 and D2 rectify the low AC voltage. Capacitor C1 smooths the pulsating DC voltage which is used to power the audio amplifier. Resistor R1 limits the current flow through light-emitting diode D3 which functions as a "power-on" indicator. The output of the three-terminal voltage regulator U1 is five volts. The five volts powers the oscillator and tremolo circuits. Capacitor C2 improves the transient response of the three-terminal voltage regulator, U1.

The monophonic electronic organ is shown in *Figure 16-2*. The frequency of oscillation of the astable multivibrator U2 is determined by the timing capacitor C3 and the timing resistor(s) selected by switches S2-S26. These normally-open pushbutton switches can select any resistor(s) in the resistor chain R7-R31. Resistors R7-R31 have been selected for the closest possible approximation of the musical notes, within one or two Hertz. The reader may choose to replace each resistor in the resistor chain with the next lowest standard value resistor in series with a 1000 ohm trimmer potentiometer.

The output of U2 is fed via volume control R5 to the input of U3 which is a TDA-2002 audio amplifier. The TDA-2002 has internal thermal overload and short circuit protection circuits. The frequency response should exceed 40 Hz to 15 kHz within three decibels. The maxi-

Note	Frequency (Hz)	Time Period (Millisec)
C	262	3.82
C#	277	3.61
D	294	3.4
D#	311	3.22
E	330	3.03
F	349	2.87
F#	370	2.7
G	392	2.55
G#	415	2.41
A	440	2.27
A#	466	2.15
B	494	2.02
C	523	1.91
C#	554	1.81
D	587	1.7
D#	622	1.61
E	659	1.52
F	698	1.43
F#	740	1.35
G	784	1.28
G#	831	1.2
A	880	1.14
A#	932	1.07
B	988	1.01
C	1,047	0.96

Table 16-1. *Notes available on both electronic organs.*

mum power supply voltage is fifteen volts. This power amplifier can deliver at least five watts into a load impedance of 2 to 16 ohms with less than five percent harmonic distortion.

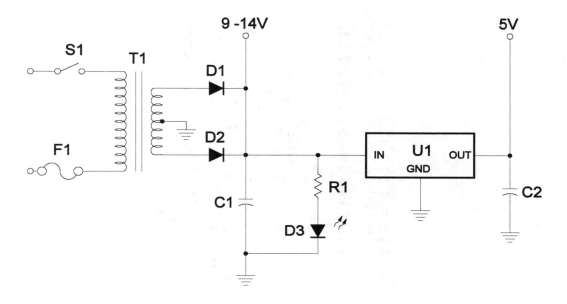

***Figure 16-1.** Power supply for both organs.*

The astable multivibrators, U2 and U4 are powered by the five volt supply while the audio amplifier U3 is powered by the nine to fourteen volt supply. If the astable multivibrators are powered by the higher voltage, the output of the astable multivibrator U2 would saturate the audio amplifier U3, which is configured as a non-inverting amplifier with a voltage gain of two.

The tremolo circuit consists of U4 and its associated components, which is configured as an astable multivibrator. The tremolo circuit oscillates at about 5 Hz. Pin 6 is connected to pin 3 (output) via resistor R2, instead of pin 7 (discharge), allowing the trigger and threshold inputs of U4 to float, instead of being tied to the power supply via a pull-up resistor. The discharge and output of the LM555 timer serve similar functions as shown in *Figure 16-3*. The tremolo can be varied with tremolo control R33 and switch S27. If S27 is opened, there is no tremolo because the tremolo circuit no longer modulates U2. When the tremolo is active, the pitch of the organ can be adjusted by pitch control R35.

The polyphonic electronic organ is shown in *Figure 16-4*. It operates similar to the monophonic organ. The main difference is that each musical note is generated by its own LM555 astable multivibrator, U2, U6-U29 and their associated components. Timing capacitor C3 is

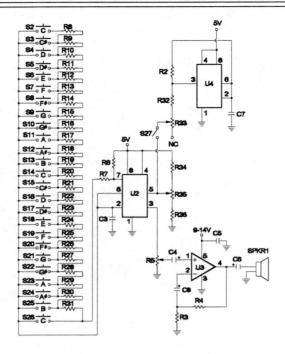

Figure 16-2. Electronic organ for playing one note at a time.

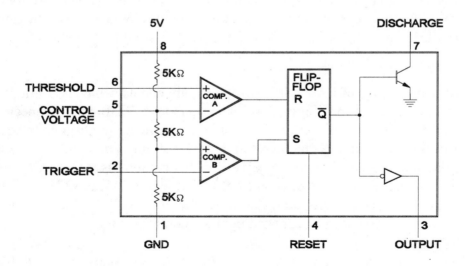

Figure 16-3. Inside the LM555 timer integrated circuit.

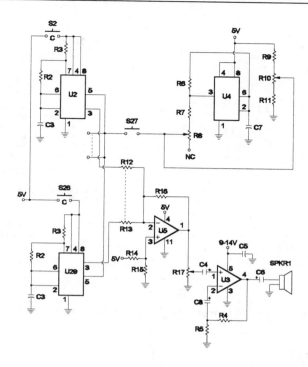

Figure 16-4. Organ for playing several notes at a time.

the same for all of the musical notes. However, timing resistor R2 is different for each of the musical notes. You may want to substitute for each R2, the next lowest standard value resistor in series with a trimmer potentiometer. The outputs of U2 and U6-U29 are fed to U5 which is configured as a unity gain summing amplifier. The tremolo circuit consists of U4 and its associated components. The audio amplifier consists of U3 and its associated components.

You may power the musical note generator circuits, tremolo and mixer circuits with the power supply used to power the audio amplifier if and only if R16 is changed from a 10,000 ohm resistor to a 1000 ohm resistor. The audio amplifier will not be saturated by the output of the summing amplifier because it now operates as a summing attenuator with an attenuation factor of ten.

If you have a doorbell transformer with a secondary rating of ten volts, the power supply shown in *Figure 16-5* may be used to power either electronic organ. The rectifier bridge consists of diodes D1, D2, D4 and D5.

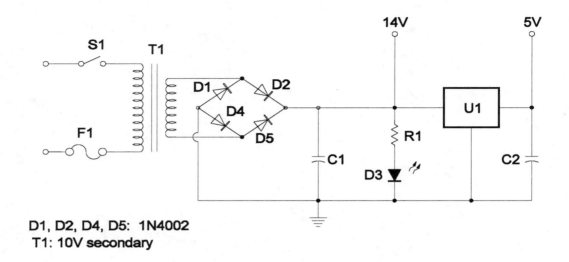

D1, D2, D4, D5: 1N4002
T1: 10V secondary

Figure 16-5. Alternate power supply for both electronic organs.

Construction

The electronic organs may be built on one or more pieces of perf board. The author used one piece of perf board for the timing resistor(s) chain and a second piece of perf board for the rest of the monophonic organ. The power supply was built on a third piece of perf board.

The timing resistors R7-R31 may each be substituted with the next lowest standard value resistor in series with a 1000 ohm trimmer potentiometer. The frequency of each musical note of the monophonic organ may therefore be accurately tuned by adjusting the appropriate trimmer potentiometer.

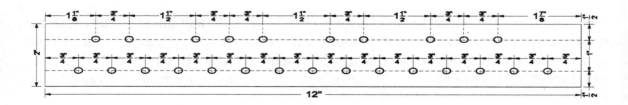

Figure 16-6. Template for keyboard, drawn one-half scale.

All Resistors Are 1/4W At 5% Unless Otherwise Noted	All Capacitors Are Rated At 25 Volts or More
R1: 1k	C1: 4700 uF
R2: 68k	C2: 0.1 uF
R3, R4: 220	C3: 56 nF
R5, R33: 10k potentiometer	C4: 10 uF
R6: 10k	C5: 68 nF
R7: 7.5k	C6: 1000 uF
R8, R9: 2.7k	C7: 1 uF
R10, R11: 2.4k	C8: 100 uF
R12, R14: 2k	U1: LM7805
R13, R34: 2.2k	U2, U4: LM555
R15 - R17: 1.8k	U3: TDA - 2002 or LM383
R18 - R21, R23: 1.5k	D1, D2: 1N4001
R22, R25 - R27: 1k	D3: LED
R24: 1.2k	S1, S27: SPST
R28, R29: 820	S2 - S26: N.O. SPST
R30, R31: 680	T1: 12.6 VCT secondary
R32: 3.3k	F1: 1/2A slow-blow
R35: 5k potentiometer	SPKR1: 4 - 8 ohm speaker
R36: 6.8k	

Table 16-2. Parts list for the monophobic electronic organ.

An adequate heat sink should be fabricated for the audio amplifier. The heat sink can be made of steel or aluminum and it should be electrically, but not thermally, isolated from the audio amplifier integrated circuit, U3.

Care should be taken to avoid cold solder joints and solder bridges. All diodes, capacitors and integrated circuits must be properly oriented.

All Resistors Are 1/4W At 5% Unless Otherwise Noted	All Capacitors Are Rated At 25 Volts or Greater
R1: 1k	C1: 4700 uF
R2: See Table 16-4	C2: 0.1 uF
R3: 10k	C3: 56 nF
R4, R5: 220	C4: 10 uF
R6: 68k	C5: 68 nF
R7: 3.3k	C6: 1000 uF
R8, R17: 10k potentiometer	C7: 1 uF
R9: 2.2k	C8: 100 uF
R10: 5k trimmer potentiometer	D1, D2: 1N4001
R11: 6.8k	D3: LED
R12 - R16: 10k	U1: LM7805
U2, U4, U6 - U29: LM555	U3: TDA - 2002 or LM383
U5: LM324	T1: 12.6 VCT secondary
S1, S27: SPST	S2 - S26: N.O. SPST
F1: 1/2A slow-blow	SPKR1: 4 - 8 ohm speaker

Table 16-3. Parts list for the polyphonic electronic organ.

The cabinet design is only limited by the reader's imagination. The author used twenty-five normally-open SPST pushbutton switches for the keyboard. A drilling template for the keyboard is shown in *Figure 16-6*. The holes drilled should be approximately one-quarter inch in diameter to suit the pushbuttons used.

The parts list for the monophonic electronic organ is given in *Table 16-2*. The parts lists for the polyphonic organ are given in *Table 16-3* and *Table 16-4*.

If the polyphonic electronic organ is built, several pieces of perf board are required to build the twenty-five musical note generating astable multivibrators. The values of R2 for the polyphonic organ are given in *Table 16-4*. The next lowest standard value resistor in series

Note	Frequency (Hz)	R2 (ohm) of Figure 16-4
C	1047	7.5k
B	988	8.2k
A#	932	8.9k
A	880	9.7k
G#	831	10.5k
G	784	11.5k
F#	740	12.5k
F	698	13.5k
E	659	14.7k
D#	622	16.2k
D	587	17.2k
C#	554	18.7k
C	523	20.2k
B	494	21.7k
A#	466	23.2k
A	440	25.0k
G#	415	26.8k
G	392	28.6k
F#	370	30.6k
F	349	30.8k
E	330	32.8k
D#	311	35.2k
D	294	37.6k
C#	277	40.3k
C	262	43.0k

Table 16-4. Resistor R2 values for the polyphonic organ.

with a trimmer potentiometer may be substituted for R2. The trimmer potentiometer is used to tune each musical note. From C (1047 Hz) to G# (831), a 1000 ohm trimmer potentiometer may be used. From G (784) to B (494), a 5000 ohm trimmer potentiometer should be used. From A# (466) to C (262), a 10,000 ohm trimmer potentiometer may be used.

The monophonic organ may be expanded to three or more octaves in two ways. The timing resistor chain may be extended to include several additional musical notes. Alternately, a second monophonic organ may be built with the resistors of the timing resistor chain selected to the next two lower or upper octaves.

The polyphonic organ may also be expanded to three or more octaves. An astable multivibrator is built and tuned to each additional musical note. The output of each additional astable multivibrator is fed via a 10,000 ohm resistor to the summing amplifier, U5, as shown in *Figure 16-4*.

Tuning and Use

If the electronic organ has been built using only fixed-value resistors, no tuning is required. Simply press the switches of the keyboard and play music.

If either electronic organ has been built using trimmer potentiometers, the organ must be tuned. Switch S27 must be in the open position. Each trimmer potentiometer is adjusted until the output of the appropriate astable multivibrator is correct according to *Table 16-1*. If a frequency counter is used to tune the organ, use the frequency data of *Table 16-1*. If an oscilloscope is used to tune the organ, use the time period data of *Table 16-1*. The organ is now ready to be played.

For tremolo, set S27 in the closed position. The tremolo and pitch controls can be adjusted for the desired sound effects.

Chapter 17

Two Radio Frequency Tone Transmitters

A radio frequency tone transmitter transmits a tone to a nearby radio receiver. An RF tone transmitter has many applications.

An RF tone transmitter can be used to keep track of your child in a busy shopping mall. A tone transmitter can be installed in stereos, televisions and automobiles. If any of these articles is stolen, the RF tone transmitter sends out a homing signal. A radio receiver is then used as a tracking device to recover the stolen article.

Two radio frequency tone transmitters are discussed. One of the radio transmitters is an amplitude modulation (AM) tone transmitter. The other radio transmitter is a frequency modulation (FM) tone transmitter.

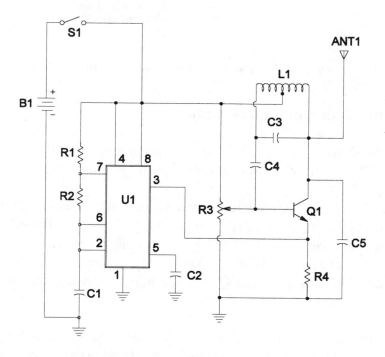

Figure 17-1. Amplitude modulation tone transmitter.

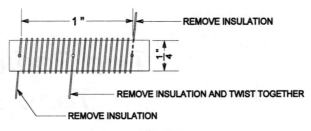

COIL CONSTRUCTION

1.) REMOVE INSULATION FOR 1/2" AT
   EACH END OF 8 FEET 30 GAUGE
   MAGNET WIRE.

2.) REMOVE 3" INSULATION AT MID-LENGTH
   OF 8 FEET 30 GAUGE MAGNET WIRE.

3.) PREPUNCH HOLES INTO 1/4" DIAMETER STRAW.

4.) SLIDE ABOUT 2" OF ONE END OF
   WIRE THROUGH AN "END" HOLE.

5.) WIND WIRE AROUND STRAW UNTIL THE
   MID-LENGTH IS REACHED.

6.) PUSH 3" OF STRIPPED WIRE THROUGH
   "CENTER" HOLE AND TWIST THGETHER.
   THIS IS THE CENTER TAP.

7.) WIND REST OF WIRE AROUND STRAW
   AND SLIDE LAST 2" THROUGH ENPTY HOLE.

Figure 17-2. Frequency modulation tone transmitter.

AM Tone Transmitter Circuit Description

The AM tone transmitter transmits a 1000 Hz tone to a nearby AM radio that is tuned to about 700 kHz. The circuit of the AM tone transmitter is shown in *Figure 17-1*.

Integrated circuit U1 and its associated components are configured as an astable multivibrator. Timing components R1, R2 and C1 determine the frequency of oscillation of U1. The output of U1 amplitude modulates transistor Q1 which is configured as a radio frequency (RF) oscillator. The tank circuit consists of L1 and C3 and it determines the frequency of oscillation of Q1. The DC bias of transistor Q1 may be adjusted by potentiometer R3. The only non-standard part of the circuit is coil L1.

If the AM tone transmitter operates at the same frequency as a local radio station, capacitor C3 may be changed. If it is replaced by a 0-to-365 pF variable capacitor, the tone transmitter may be tuned to any "blank" spot on the AM radio band.

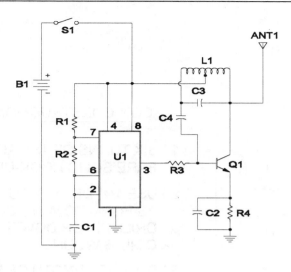

Figure 17-3. Construction of the coil for the FM tone transmitter.

The details of the construction of coil L1 are shown in *Figure 17-2*. The antenna ANT1 can be a piece of wire not to exceed nine and three-quarter feet in length. Battery B1 can be in the range of three to nine volts.

FCC regulation 15.113 prohibits the unlicensed use of an RF transmitter where the power input to the final radio stage exceeds 100 milliwatts. If battery B1 is greater than three volts, resistor R4 can be increased to ensure that the AM tone transmitter meets the requirements of the FCC.

FM Tone Transmitter Circuit Description

The FM tone transmitter transmits an 875 Hz tone to a nearby FM radio that is tuned to about 98 MHz. The circuit of the FM tone transmitter is shown in *Figure 17-3*.

Integrated circuit U1 and its associated components are configured as an astable multivibrator. Timing components R1, R2 and C1 determine the frequency of oscillation of U1. The output of U1 frequency modulates transistor Q1 which is configured as a radio frequency (RF) oscillator. The tank circuit consists of L1 and C3 and it determines the frequency of oscillation of Q1. The base drive for transistor Q1 is supplied by resistor R3. The only non-standard part of the circuit is coil L1.

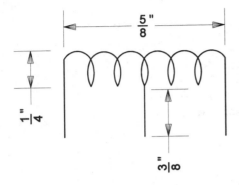

COIL CONSTRUCTION

1.) SIX TURNS OF 18 GAUGE
   BARE SOLID HOOK-UP WIRE.

2.) USE 1/4" DRILL BIT OR DOWEL
   AS FORM TO WIND COIL. REMOVE
   DRILL BIT OR DOWEL WHEN
   COIL IS WOUND.

3.) SOLDER LENGTH OF HOOK-UP
   WIRE AT MIDPOINT TO SERVE AS
   THE CENTER TAP.

Figure 17-4. Construction of the coil for the FM tone transmitter.

If the FM tone transmitter operates at the same frequency as a local radio station, capacitor C3 may be changed. If it is replaced by a 15-to-60 pF variable capacitor, the FM tone transmitter can be tuned to a "blank" spot on the FM radio band.

The details of the construction of coil L1 are shown in *Figure 17-4*. The antenna ANT1 can be a piece of wire not to exceed several inches in length. Battery B1 can be in the range of three to nine volts.

FCC regulation 15.113 prohibits the unlicensed use of an RF transmitter where the power input to the final radio stage exceeds 100 milliwatts. If battery B1 is greater than three volts, resistor R4 can be increased to ensure that the AM tone transmitter meets the requirements of the FCC.

Construction and Use of the AM Tone Transmitter

The AM tone transmitter can be built on a piece of perf board. The integrated circuit and transistor must be properly oriented. Care should be taken to avoid solder bridges or cold solder joints. The parts list for the AM tone transmitter is given in *Table 17-1*.

All Resistors Are 1/4W At 5% Unless Otherwise Noted
R1: 1k
R2: 15k
R3: 50k potentiometer
R4: 100
C1: 47 nF
C2: 0.01 uF
C3: 470 pF
C4: 0.1 uF
C5: 0.005 uF
L1: See text
U1: LM555 timer
Q1: 2N2222
B1: See text
ANT1: See text
S1: SPST

Table 17-1. Parts list for the AM tone transmitter.

The details for the construction of coil L1 are shown in *Figure 17-2*. The air core of the coil is a one-quarter inch diameter drinking straw. Eight feet of 30 gauge magnet wire is used for the windings of the coil. Remove one-half inch of insulation from each end of the wire. Remove three or four inches of insulation from the mid-length of the 30 gauge magnet wire. Make three sets of holes about one-half inch apart from each other. Insert two inches of one end of the magnet wire into one of the "end" holes as shown in *Figure 17-2*. Wind the wire in a clockwise direction around the straw. When the stripped mid-length of the wire is reached, insert it into the "middle" hole and make a loop. This is the center tap of the coil. Wind the rest of the wire also in a clockwise direction and pass the last two inches of the wire through the empty "end" hole.

Tune an AM radio to around 700 kHz. Turn on the AM tone transmitter. A 1000 Hz tone should be heard on the radio. If not, tune the AM radio until a tone is heard. Adjust potenti-

All Resistors Are 1/4W At 5% Unless Otherwise Noted
R1, R3: 15k
R2: 75k
R4: 100
C1, C2: 0.01 uF
C3: 47 pF
C4: 39 pF
L1: See text
U1: LM555 timer
Q1: 2N2222
B1: See text
ANT1: See text
S1: SPST

Table 17-2. Parts list for the FM tone transmitter.

ometer R3 for a clear sounding tone. The transmission range is about twenty-five feet when the antenna ANT1 is nine and three-quarter feet long. The AM tone transmitter must not violate FCC regulation 15.113.

An ammeter should be used to measure the current drawn by the AM tone transmitter. This current multiplied by the battery voltage B1 should not exceed 100 milliwatts. If it does, either decrease the battery voltage B1 or increase resistor R4.

Construction and Use of the FM Tone Transmitter

The FM tone transmitter can be built on a piece of perf board. The integrated circuit and transistor must be properly oriented. Care should be taken to avoid solder bridges or cold solder joints. The parts list for the FM tone transmitter is given in *Table 17-2*.

The details for the construction of coil L1 are shown in *Figure 17-4*. Six turns of 18 gauge solid-core bare hookup wires used to wind coil L1. The coil is wound on a one-quarter inch diameter drill bit or wooden dowel. An extra one-quarter to one-half inch of wire is left at each end. Remove the drill bit or wooden dowel once the coil is wound. Solder a length of 18 gauge solid-core bare hookup wire to the mid-length of the coil. This is the center tap of the coil.

Tune an FM radio to around 98 MHz. Turn on the FM tone transmitter. An 875 Hz tone should be heard on the radio. If not, tune the FM radio until a tone is heard. The transmission range is about one hundred feet when the antenna ANT1 is seven inches long. The FM tone transmitter must not violate FCC regulation 15.113.

An ammeter should be used to measure the current drawn by the FM tone transmitter. This current multiplied by the battery voltage B1 should not exceed 100 milliwatts. If it does, either decrease the battery voltage B1 or increase resistor R4.

Chapter 18

Two FM Voice Transmitters

An FM voice transmitter has many uses. It can be used as a baby monitor, as a wireless intercom, and as a voice broadcast transmitter.

Two FM voice transmitter projects are described in this chapter. The first project is designed for use with an electret mike element. The second project is designed for use with a microphone or any other suitable signal source.

Both projects feature an operational-amplifier front end with gain and modulation controls. The oscillator stage is a simple one-transistor tunable oscillator.

Both projects operate in the FM band; that is, they operate in the range of 88 MHz to 108 MHz. All wire leads should be as short as possible to eliminate stray lead capacitances.

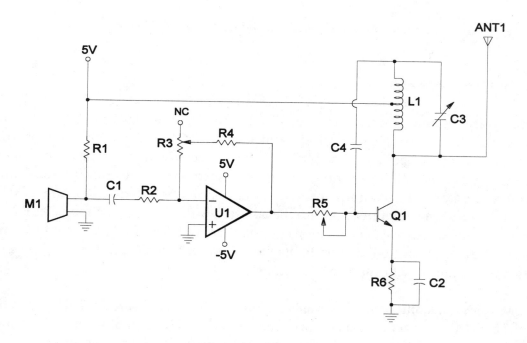

Figure 18-1. *FM voice transmitter with electret mike element.*

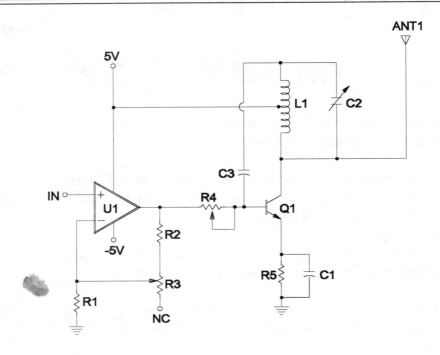

Figure 18-2. FM voice transmitter for microphone input.

Circuit Descriptions

An FM voice transmitter with an electret mike element is shown in *Figure 18-1*. Resistor R1 provides DC bias to the electret mike element. Capacitor C1 decouples the DC bias from the negative input of the operational-amplifier U1 which is configured as an inverting amplifier. The input impedance of the inverting amplifier is determined by resistor R2. Resistor R2 is therefore selected to be ten times the output impedance of electret mike element M1. The gain of the inverting amplifier stage can be adjusted with trimmer potentiometer R3. Trimmer potentiometer R5 is the modulation control. Resistor R6 limits the current flow through transistor Q1. Resistor R6 therefore limits the RF power transmitted. The frequency of operation can be tuned with variable capacitor C3.

An FM voice transmitter with a microphone input is shown in *Figure 18-2*. The operational-amplifier U1 is configured as a non-inverting amplifier. The gain of the non-inverting amplifier stage can be adjusted with trimmer potentiometer R3. The input impedance of the non-inverting amplifier is high; therefore, any suitable signal source may be used for this

All Resistors Are 1/4W At 5% Unless Otherwise Noted
R1: 1k
R2, R4: 10k
R3: 2M trimmer potentiometer
R5: 10k trimmer potentiometer
R6: 100 ohm @ 1/2W
C1, C2: 0.01 uF
C3: 15 - 60 pF variable
C4: 39 pF
L1: See text
ANT1: See text
Q1: 2N2222
U1: LM308
M1: Electret mike element, Radio Shack #270-090

Table 18-1. Parts list for electret mike element FM transmitter.

project. Trimmer potentiometer R4 is the modulation control. Resistor R5 limits the current flow through transistor Q1. Resistor R5 therefore limits the RF power transmitted. The frequency of operation can be tuned with variable capacitor C2.

Both projects should operate properly if proper construction techniques are employed.

Construction

Both FM voice transmitters can be built on a piece of perf board. The parts list for the FM transmitter with the electret mike element input is given in *Table 18-1*. The parts list for the FM transmitter with the microphone input is given in *Table 18-2*.

All Resistors Are 1/4W At 5% Unless Otherwise Noted
R1, R2: 1k
R3: 500k trimmer potentiometer
R4: 10k trimmer potentiometer
R5: 100 ohm @ 1/2W
C1: 0.01 uF
C2: 15 - 60 pF variable
C3: 39 pF
U1: LM308
Q1: 2N2222
L1: See text
ANT1: See text

Table 18-2. Parts list for microphone input FM transmitter.

Point-to-point wiring should be used and all leads should be kept as short as possible, preferably less than one inch. Care should be taken to avoid cold solder joints or solder bridges. If poor construction techniques are used, the circuit will not operate properly and it will be susceptible to stray lead and body capacitances.

The antenna can either be a piece of wire or a replacement telescoping FM antenna. Coil L1 is the same coil that was used in the FM tone transmitter described in the previous chapter. It consists of six turns of 18 gauge solid bare hookup wire wound around a one-quarter inch form. Once the coil is wound, the form is removed and a center tap is securely crimped and soldered to the center of the coil. The coil is shown in *Figure 18-3*.

Testing and Use

The FM transmitter should function properly the first time. The gain and modulation controls are adjusted until a clear voice is heard on the FM radio. The variable capacitor is used to tune the FM voice transmitter to a blank spot on the FM dial.

If the FM voice transmitter is not functioning properly, check for cold solder joints or solder bridges. At these high frequencies, cold solder joints affect the operation of the circuit by presenting unwanted capacitances. The operational-amplifier and the transistor must be properly oriented.

Verify that the transistor oscillator is receiving a modulating signal with an oscilloscope. The vocal signal should be present at the output of the operational-amplifier and at the base and emitter leads of the transistor.

If the circuit does not oscillate, verify that the center tap is firmly crimped and soldered to the coil. The FM voice transmitter should be housed in a metal box to shield against stray lead and body capacitances.

The FM voice transmitter should provide years of trouble free service as an FM baby monitor, as an FM wireless intercom, or as an FM voice broadcast transmitter.

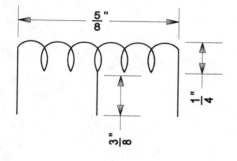

COIL CONSTRUCTION

1.) SIX TURNS OF 18 GAUGE BARE SOLID HOOK-UP WIRE.

2.) USE 1/4" FORM TO WIND COIL. REMOVE FORM WHEN COIL IS WOUND.

3.) CRIMP FIRMLY AND SOLDER LENGTH OF 18 GAUGE BARE SOLID HOOK-UP WIRE AT MIDPOINT TO SERVE AS THE CENTER TAP.

Figure 18-3. Details of FM voice transmitter coil construction.

Appendix

Problem Solutions

Chapter 1 Problem Solutions

Problem 1-1. An amplifier requires an input signal. An oscillator does not require an input signal. Most amplifiers have negative feedback circuits. Oscillators have positive or regenerative feedback circuits. An amplifier produces an amplified version of its input signal at its output. An oscillator produces an output signal with a definite shape and frequency.

Problem 1-2. At frequencies below approximately 100 kHz, inductors are large with non-ideal characteristics.

Problem 1-3. An electromechanical oscillator generates an output with a very stable frequency. The frequency reference is a mechanical device with a constant source of vibrations.

Problem 1-4. The crystal oscillator is an electromechanical oscillator. Other types of electromechanical oscillators are the metal bar oscillator, the rod oscillator and the tuning fork oscillator.

Chapter 2 Problem Solutions

Problem 2-1. Negative feedback reduces the gain sensitivity to component changes, reduces nonlinear distortion, reduces the effect of noise, increases or decreases the input and output resistances, increases the bandwidth, reduces the gain and stabilizes the circuit.

Problem 2-2. Let $A = 10$ and $B = 0.1$. Hence, $AB = 1$, $A' = A/(1 + AB) = 5$, and $A = 1/B = 10$.
Let $A = 10$ and $B = 0.01$. Hence, $AB = 0.1$, $A' = A/(1 + AB) = 9.1$, and $A = 1/B = 100$.
Let $A = 100$ and $B = 0.01$. Hence, $AB = 1$, $A' = A/(1 + AB) = 50$, and $A' = 1/B = 100$.

Let A = 1000 and B = 0.01. Hence, AB = 10, A' = A/(1 + AB) = 90.9, and
A' = 1/B = 100.
Let A = 1000 and B = 0.1. Hence, AB = 100, A' = A/(1 + AB) = 9.9, and
A' = 1/B = 10.

Problem 2-3. A'm = Am/(1 + AmB) = 9.9

Problem 2-4. w_{HF} = w_H(1 + AmB) = 1.01 MHz

Problem 2-5. w_{LF} = w_L/(1 + AmB) = 0.99 Hz

Problem 2-6. A' = A/(1 + AB) = 9.1, R'$_{IN}$ = R$_{IN}$(1 + AB) = 110 kohm, R'$_O$ = R$_O$/(1 + AB) = 9.1 ohm

Problem 2-7. A' = A/(1 + AB) = 9.1, R'$_{IN}$ = R$_{IN}$/(1 + AB) = 909 ohm, R'$_O$ = R$_O$(1 + AB) = 1100

Problem 2-8. A' = A/(1 + AB) = 9.1, R'$_{IN}$ = R$_{IN}$(1 + AB) = 110 kohm, R'$_O$ = R$_O$(1 + AB) = 1100

Problem 2-9. A' = A/(1 + AB) = 9.1, R'$_{IN}$ = R$_{IN}$/(1 + AB) = 909 ohm, R'$_O$ = R$_O$/(1 + AB)=9.1

Problem 2-10. A voltage amplifier.

Problem 2-11. A current amplifier.

Problem 2-12. A transconductance amplifier.

Problem 2-13. A transresistance amplifier.

Problem 2-14. A' = A/(1 - AB) = 10/0 = infinity

Problem 2-15. A' = 10 when B = 0.0, and A' = 25 when B = 0.06.

Problem 2-16. Positive feedback increases the gain of a circuit, destabilizes the circuit, and increases the distortion content of the output signal of a circuit.

Problem 2-17. The poles of a stable circuit lie in the left half of the s-plane.

Problem 2-18. The gain margin is the difference between unity and the magnitude of the loop gain of an amplifier at a phase angle of 180 degrees.

Problem 2-19. The phase margin is the difference between 180 degrees and the phase angle at which the amplifier loop gain is unity.

Problem 2-20. An amplifier is considered stable when the point (-1,0) is to the right of the intersection of the negative real axis at frequency w_{180}.

Problem 2-21. An amplifier is considered stable when its loop gain magnitude is less than unity at a frequency of less than 180 degrees phase shift. It is also considered stable when its loop gain magnitude at w_{180} is less than unity.

Chapter 3 Problem Solutions

Problem 3-1. Two types of oscillators are the tuned inductive-capacitance (LC) type and the relaxation resistance-capacitance (RC) type.

Problem 3-2. An electromechanical oscillator uses a constant source of vibration to yield a stable frequency output. A crystal oscillator is an electromechanical oscillator.

Problem 3-3. At the frequency of oscillation the magnitude of the loop gain is unity and the phase of the loop gain is zero.

Problem 3-4. In theory-an impulse. In practice-spurious signals.

Problem 3-5. Resistors generate thermal noise. Transistors generate shot noise.

Problem 3-6. A noise signal can be considered as a series of impulses.

Problem 3-7. The poles are on the imaginary axis or on the right half of the s-plane.

Problem 3-8. The component values of the feedback network and the open-loop gain of the amplifier determine the pole positions of an oscillator.

Problem 3-9. Two important oscillator specifications are the frequency of oscillation and the condition required for sustained oscillation.

Problem 3-10. The frequency of oscillation is determined from the imaginary part of the poles of the transfer function. The real part of the poles of the transfer function yields the condition required for sustained oscillation.

Problem 3-11. $w_0 = 1/RC = 10,000$ rad/sec. $f_0 = w_0/2 * PI = 1592$ Hz.

Problem 3-12. The phase-shift oscillator requires an open-loop gain of at least 29 for sustained oscillation.

Problem 3-13. $w_0 = 1/SQRT(6)RC = 4080$ rad/sec. $f_0 = w_0/2 * PI = 650$ Hz.

Problem 3-14. $w_0 = 1/RC = 10,000$ rad/sec. $f_0 = 1592$ Hz.

Problem 3-15. They all have LC tuned parallel circuits. The Colpitts oscillator has no tap on the oscillator coil. The Hartley oscillator has a tap on the oscillator coil. The Clapp oscillator has no tap on the oscillator coil but it has an additional capacitor in series with the oscillator coil. The Armstrong oscillator has two separate coils in its tank circuit.

Problem 3-16. $w_0 = 1/SQRT(LC) = 316228$ rad/sec. $f_0 = 50,355$ Hz.

Problem 3-17. $w_0 = 1/SQRT(LC) = 316228$ rad/sec. $f_0 = 50,355$ Hz.

Problem 3-18. The Armstrong oscillator is used as a local oscillator in receivers, as a signal source in signal generators and as a variable-frequency oscillator in medium and high frequency ranges.

Problem 3-19. The piezoelectric effect is the interrelation of mechanical and electrical stresses in crystals. A crystalline structure changes shape when an EMF is applied to it and it generates an EMF when its undergoes mechanical stress.

Problem 3-20. The series or motional elements; that is, the series inductance and capacitance, determine the series resonant frequency of a crystal.

Problem 3-21. The series resistance determines the Q factor of the resonator.

Problem 3-22. w_o = 1/SQRT(L1C1) = 10,000,000 rad/sec. f_o = 1.6 MHz.

Problem 3-23. The series capacitance is specified when the crystal is pulled.

Problem 3-24. The shunt capacitance is in parallel with the series or motional elements. It is a function of the crystal holder. The shunt capacitance is also called the parallel capacitance.

Problem 3-25. The crystal operates in either its fundamental mode or its overtone mode.

Problem 3-26. The crystal oscillates at its natural resonant frequency in its fundamental mode. The fundamental frequency depends on the crystal material, how it was cut and temperature.

Problem 3-27. The crystal oscillates at an integer multiple of its natural frequency in its overtone mode. This frequency is not a harmonic of its natural frequency.

Problem 3-28. Rochelle salts are used in microphones and phonograph pickups.

Problem 3-29. Quartz crystals are used in radio-frequency (RF) oscillators.

Problem 3-30. The Pierce oscillator is used because different crystals can be switched in and out without having to retune the oscillator circuit.

Problem 3-31. The frequency, amplitude and output distortion are controlled independently. The Q factor of the bandpass filter determines the amount of distortion in the output.

Problem 3-32. w_o = 1/RC = 10,000 rad/sec. f_o = 1592 Hz.

Problem 3-33. The output of an operational-amplifier twin-T oscillator is distorted if the potentiometer resistance is too low.

Problem 3-34. There is no output from an operational-amplifier twin-T oscillator if the potentiometer resistance is too high.

Chapter 4 Problem Solutions

Problem 4-1. The amplifier can be protected by a fuse, a resistor or a panel light in the DC supply line.

Problem 4-2. Varying power supply, temperature changes and mechanical contraction and expansion.

Problem 4-3. A voltage regulator and a temperature-compensating capacitor in parallel with the tank circuit can be used to prevent output fluctuations.

Problem 4-4. A high Q factor, a low L/C ratio and a light load contribute to the stability of an oscillator.

Problem 4-5. Parasitic oscillations are unwanted spurious oscillations.

Problem 4-6. The inductance and capacitance of leads cause parasitic oscillations.

Problem 4-7. Parasitic oscillations can be eliminated by placing a small resistance (10-100 ohms) in series with the offending lead.

Problem 4-8. The loop gain of an oscillator is kept at unity by a nonlinear gain control circuit.

Problem 4-9. Two types of gain control circuits are the limiter and the circuit parameter control.

Problem 4-10. Let $C1 = C2 = C3 = 0.047$ uF. $R = R1 = R2 = 2,765$ ohms. $R3 = 3,900$ ohms. $R4 = 3,900$ ohms. $R_I = 8,775$ ohms. $R_F = 246$ kohms.

Problem 4-11. Let $C = C1 = C2 = 0.047$ uF. $D1 = D2 = 1N4735A$. $R = R1 = 6,773$ ohms. $R2 < R$; therefore, let $R2 = 6,400$ ohms.

Problem 4-12. Let $C = C1 = C2 = 0.047$ uF. $C3 = 0.094$ uF. $R = R1 = R2 = 6,776$ ohms. $R3 = 1,694$ ohms. $R4 = 3,388$ ohms. $R5 = 13,552$ ohms.

Chapter 5 Problem Solutions

Problem 5-1. The resonant frequency of a tuned circuit determines the period of the cycle of a sinusoidal oscillator.

Problem 5-2. The charge and discharge times of a capacitor determine the period of the cycle of a non-sinusoidal oscillator.

Problem 5-3. The circuit triggers. The output voltage increases to Vcc. As Vin increases, V_{B2} decreases, causing V_{BE2} to decrease. Also B2 as Vin decreases, V_E increases, also causing V_{BE2} to decrease. The change in V_E turns off Q2. The switching is regenerative. The action is similar for Q1.

Problem 5-4. $V_{B2} = R2Vcc/(R1 + R2 + R_{C1}) = 3000 \times 15/(3000 + 3000 + 1500) = 45/7.5 = 6$ V. Transistor Q1 is off.

Problem 5-5. $V_{E2} = (V_{B2} - V_{BE2}) = 6 - 0.6 = 5.4V$

Problem 5-6. $I_{C2} = I_{E2} = V_{E2}/R_E = 5.4/1500 = 3.6$ mA
$Vo = Vcc - I_{C2}R_{C2} = 15 - 0.0036 \times 1000 = 15 - 3.6 = 11.4$ V

Problem 5-7. $V_{CE2} = Vo - V_E = 11.4 - 5.4 = 6V$; therefore, Q2 is not in saturation.

Problem 5-8. $Vin = V_E + V_{BE1} = 5.4 + 0.7 = 6.1V$. Q1 turns on and Q2 turns off.

Problem 5-9. The turn on voltage for Q1 is not affected by temperature or transistor replacement because V_{BE} active tends to cancel the BE cut-in voltage giving $Vin = V_{B2} - 0.1$.

Problem 5-10. Voltage V_{B2} is lower when Q1 turns on. Vin, V_E and V_{B2} are B2 approximately equal before Q1 turns on. If V_{B2} decreases and V_E stays the same, then Q2 must turn off. If Vin decreases, V_E decreases, causing V_{BE} to increase.

Problem 5-11. $V_{B2} = I_{B2} \times R1 = I_{B2} \times 3000$
$V_E = V_{B2} - V_{BE2} = V_{B2} - 0.7$ for Q2 turning on.
$V_E = I_{C1} \times R_E = I_{C1} \times 1500$
$V_E = V_{B2} - 0.7$ can be written as: $1500I_{C1} = 3000I_{B2} - 0.7(1)$

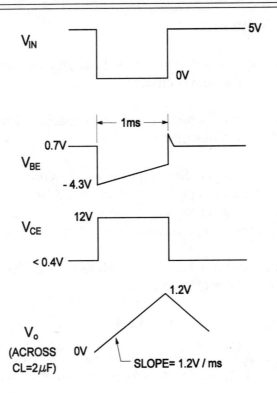

Figure P6-1. *Waveforms for Problem 6-5*.

Using Kirchoff's voltage law: $(R1 + R2)I_{B2} + (I_{C1} + I_{B2})R_{C1} = Vcc(2)$
$(3000 + 3000)I_{B2} + (I_{C1} + I_{B2})1500 = 15$
$6000I_{B2} + 1500I_{C1} + 1500I_{B2} = 15$
$7500I_{B2} + 1500I_{C1} = 15(3)$
Substituting equation (1) into equation (3):
$7500I_{B2} + 3000I_{B2} - 0.7 = 15$
$10,500I_{B2} = 15.7$
$I_{B2} = 15.7/10,500 = 1.495$ mA
$V_{B2} = I_{B2} \times R1 = 0.001495 \times 3000 = 4.49$V.

Problem 5 - 12. $V_E = V_{B2} - V_{BE2} = 4.49 - 0.7 = 3.79$V
$\qquad V_{B1} = V_{B2} + V_{CE1}(sat) = 4.49 + 0.1 = 4.59$V

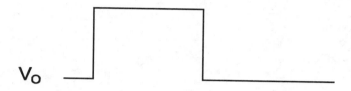

Figure P6-2. *Waveforms of a monostable multivibrator.*

Chapter 6 Problem Solutions

Problem 6-1. $I_C(sat) = (Vcc - V_{CE}(sat))/Rc = (6 - 0.4)/600 = 5.6/600 = 9.3$ mA.
Let $h_{FE}(min) = 50$; therefore, $I_B(sat) = I(sat)/h_{FE} = 0.093/50 = 186$ uA.
$R_B = (V_{IN} - V_{BE})/I_B(sat) = (5 - 0.7)/0.000186 = 4.3/0.000186 = 23,118$ ohms.
At -55 degrees Celsius, $I_B(sat) = 0.000186/0.58 = 32$ uA.
At -55 degrees Celsius, $R_B = 4.3 \times 0.58/0.000186 = 13,409$ ohms.

Problem 6-2. $I_C(sat) = (6 - 0.4)/6000 = 0.9$ mA. Let $h_{FE} = 50 \times 0.7 = 35$
$I_B(sat) = I_C(sat)/h_{FE} = 0.9/35 = 25.7$ uA and $h_{FE}(-55$ deg.$) = 0.4 \times 50 = 20$.
At -55 degrees Celsius $I_B(sat) = 0.9/20 = 45$ uA.
$R_B = (V_{IN} - V_{BE})/I_B = (5 - 0.7)/0.000045 = 4.3/0.000045 = 95,556$ ohms.

Problem 6-3. The storage time is increased.

Problem 6-4. The turn-on time is increased.

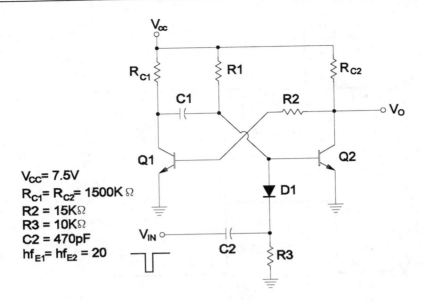

$V_{CC} = 7.5V$
$R_{C1} = R_{C2} = 1500K\,\Omega$
$R2 = 15K\Omega$
$R3 = 10K\Omega$
$C2 = 470pF$
$hf_{E1} = hf_{E2} = 20$

Figure P6-3. *Monostable multivibrator circuit.*

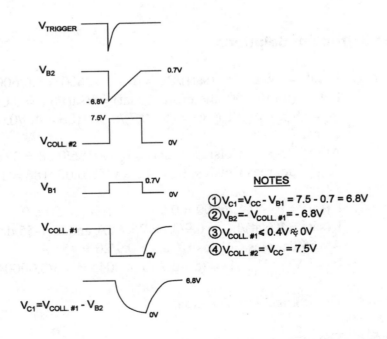

NOTES

① $V_{C1} = V_{CC} - V_{B1} = 7.5 - 0.7 = 6.8V$

② $V_{B2} = -V_{COLL. \#1} = -6.8V$

③ $V_{COLL. \#1} \leq 0.4V \approx 0V$

④ $V_{COLL. \#2} = V_{CC} = 7.5V$

Figure P6-4. *Waveforms for Problem 6-13.*

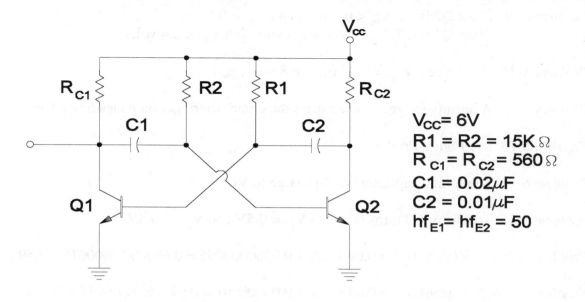

Figure P6-5. *Astable multivibrator circuit for Problem 6-18.*

Problem 6-5a. See *Figure P6-1*.

Problem 6-5b. $R_B(max) = h_{FE}(min)R_C = 40 \times 5000 = 200,000$ ohms.

Problem 6-5c. See *Figure P6-1*.

Problem 6-6. Active pull-up is using an active device to pull-up the output voltage to the power supply voltage. The active device draws negligible current until it conducts.

Problem 6-7. The monostable circuit has one stable state. It produces an output pulse each time it is triggered. The monostable multivibrator is also called a one-shot multivibrator. *Figure P6-2* shows the input and output waveforms of a typical monostable multivibrator.

Problem 6-8. A monostable multivibrator consists of an inverter and an off-gated switch as shown in *Figure P6-3*.

Problem 6-9. Transistor Q2 of *Figure P6-3* is normally ON.

Problem 6-10. Since Q2 is on, $V_{C2} < 0.4$ volts and $V_{B2} = 0.7$ volts. Since Q1 is off, $V_{C1} = Vcc = 7.5$ volts and $V_{B1} < 0.4$ volts.

Problem 6-11. $V_{C1} = Vcc - V_{BE2} = 7.5 - 0.7 = 6.8$ volts.

Problem 6-12. A negative trigger pulse causes the circuit to trigger on its leading edge.

Problem 6-13. See *Figure P6-4.*

Problem 6-14. The timing capacitor would charge to Vcc.

Problem 6-15. Transistor Q2 turns on when $V_{B2} = 0.5V$ and $V_{C1} = -0.5V$.

Problem 6-16a. $T = 0.69R1C1 = 0.69 \times 22k \times 0.0000000005 = 0.69 \times 0.00000011 = 7.59$ uS.

Problem 6-16b. $T = 0.69R1C1 = 0.69 \times 18k \times 0.00000001 = 0.69 \times 0.00000018 = 0.1242$ mS.

Problem 6-17. $R1 < h_{FE2}R_{C2} = 20 \times 1500 = 30,000$ ohms.

Problem 6-18. *Figure P6-5* shows an astable multivibrator circuit.
$T = T1 + T2 = 0.69R1(C1 + C2) = 0.69 \times 15000 \times 0.00000003 = 310.5$ uS.
Duty cycle at Q2: $DC = 0.69R2C2/T = 0.69 \times 15000 \times 0.00000001/310.5$.
$DC = 103.5$ uS/310.5 uS $= 0.333$.

Chapter 7 Problem Solutions

Problem 7-1. $B = R1/(R1 + R2) = 2200/(2200 + 100,000) = 2.2/102.2 = 0.0215$
$(1 - B) = R2/(R1 + R2) = 100,000/(2200 + 100,000) = 100/102.2 = 0.978$
Take Vo to equal Vcc when the operational amplifier is saturated.
UTL $= (1 - B)Vr + BVo = 0.978 \times 2 + 0.0215 \times 12 = 1.956 + 0.258 = 2.214$ volts.
LTL $= (1 - B)Vr - BVo = 0.978 \times 2 - 0.0215 \times 12 = 1.956 - 0.258 = 1.698$ volts.
$V = UTL - LTL = 2.214 - 1.698 = 0.516$ volts.

Problem 7-2. $t_p = 1.1R1C1 = 1.1 \times 10,000 \times 0.00000001 = 110$ uS.

Problem 7-3. $T = R1C1 * \ln[(1 + \{0.7V/Vo\})/(1 - R4/\{R3 + R4\})] = 0.0001 * \ln(1.7/0.167)$
$T = 0.0001 * \ln 10.179641 = 0.000232 = 232$ uS.

Problem 7-4. The approximation $T = 0.69R1C1$ can be used. $T = 69$ uS. If the exact formula is used, $T = 76$ uS. This proves that under the conditions that $R3 = R4$ and $Vo >> 0.7$ volts, $T = 0.69R1C1$ can be used.

Problem 7-5. $F = 1/2RC = 1/2 \times 20,000 \times 0.00000001 = 1000/0.4 = 2500$ Hz.

Problem 7-6. $F = 1/2R_tC_t = 1/2000 \times 0.000001 = 1000/2 = 500$ Hz.

Problem 7-7. $R1 > Vcc/0.2 = 15/0.2 = 75$ ohms.
$F = 1.44/(R1 + 2R2)C1 = 1.44/(10k + 9.4k)0.00000001 = 1440/0.194 = 7423$ Hz.
$DC = R2/(R1+2R2) = 4700/(10k + 9.4k) = 4700/19,400 = 0.24$

Problem 7-8. $t_p = 1.2R1C1 = 1.2 \times 22,000 \times 0.00001 = 1.2 \times 0.22 = 264$ mS.

Chapter 8 Problem Solutions

Problem 8-1. $V_p = Vcc = 10$ volts and $V_s = n_sVcc = 10/2 = 5$ volts.

Problem 8-2. $I_B = V_s/R_B = 5/1000 = 5$ mA.
$I_C = n_sI_B = 0.5 \times 5 = 2.5$ mA and $I_m = 0A$ at $t = 0$ seconds.
The transistor is saturated at this point.

Problem 8-3. $di_m/dt = di_C/dt = V_p/L_m = 10/0.001n = 10$ A/mS.

Problem 8-4. The transistor turns off regeneratively when the collector current reaches $h_{FE}I_B = 20 \times 5 = 100$ mA. The magnetizing current, i_m, FEB must increase by $(100 - 2.5) = 97.5$ mA.
$t_p = L_m \, xi_m/V_p = 0.001 \times 97.5/10 = 9.75$ uS.

Problem 8-5. The output pulse width, t_p, is dependent on the transistor p gain or h_{FE}. Since h_{FE} is affected by temperature and/or transistor replacement, the output pulse width t_p is also affected by temperature and/or transistor replacement.

Problem 8-6. $V_p + n_s V_p = Vcc$; therefore, $V_p = 12/1.5 = 8V$, $V_s = 4V$ and $V_L = 2V$.

Problem 8-7. $I_E = V_s/R_E = 4/100 = 40$ mA. The emitter current is constant.

Problem 8-8. $i_B = i_p = 0A$ and $i_m = i_C = I_E = 40$ mA.

Problem 8-9. The pulse width is determined when $i_m = I_E = 40$ mA.
$t_p = L_m I_E/V_p = 0.0016 \times 0.04/8 = 0.000064/8 = 8$ uS.

Problem 8-10. The pulse width is shortened when a resistor of 200 ohms is placed across the load winding. $I_L = V_L/R_L = 2/200 = 10$ mA. The reflected load current in the primary is $n_L I_L = 10/4 = 2.5$ mA; therefore, the magnetizing current rises to $I_E - 2.5 = 40 - 2.5 = 37.5$ mA.
$t_p = L_m i_m/V_p = 0.0016 \times 0.0375/8 = 0.00006/8 = 7.5$ uS.

Problem 8-11. Reflected load: $R_L/n^2 = 200/(0.25)^2 = 3200$ ohms.

Problem 8-12. The backswing voltage in this circuit is due to the collapse of the magnetizing current in the inductor. If the backswing voltage is left undamped, oscillation and transistor breakdown can occur.

Problem 8-13. The backswing voltage can be suppressed without affecting the pulse width by placing a diode and a damping resistor across the primary winding.

Problem 8-14. The maximum possible output frequency is lowered when a diode and a damping resistor are placed across the primary winding.

Chapter 9 Problem Solutions

Problem 9-1. A negative-resistance device has decreasing current for increasing voltage in some part of its VI-characteristic curve.

Problem 9-2. Tunnel diode, Shockley diode, SUS, SBS and diac are two-terminal negative-resistance devices.

Problem 9-3. SCR, triac, UJT and PUT are three-terminal negative-resistance devices.

Problem 9-4. The SCS is a four-terminal negative-resistance device.

Problem 9-5. The tunnel diode is voltage controllable.

Problem 9-6. The tunnel diode switches quickly.

Problem 9-7. The tunnel diode has a low signal swing, only has two terminals and is difficult to fabricate in an integrated circuit.

Problem 9-8. The neon tube is the earliest negative-resistance device.

Problem 9-9. The SUS switches quickly.

Problem 9-10. The SUS has a slow turn-off time.

Problem 9-11. The SBS is useful in AC circuits because it can be made to trigger on positive and negative half cycles of the input signal.

Problem 9-12. The SCR conducts one way. The triac is bidirectional.

Problem 9-13. The UJT and PUT are useful in relaxation oscillator circuits.

Problem 9-14. In the bistable mode, the load line cuts the VI-characteristic in each positive slope segment. In the monostable mode, the load line cuts the VI-characteristic only on one positive slope segment. In the astable mode, the load line cuts the VI-characteristic only on its negative slope segment.

Problem 9-15. Positive-feedback oscillators use active components and negative-resistance oscillators use passive components.

Problem 9-16. $t_{OFF} = R1C1 * \ln(1/[1 - n]) = 10k \times 0.001 \text{ uF} \times \ln(1/[1 - 0.65])$
$t_{OFF} = 0.00001 \times \ln(1/0.35) = 0.00001 \times \ln(2.857) = 0.0000105 \text{ sec.}$

Chapter 10 Problem Solutions

Problem 10-1. The Miller integrator and bootstrap circuits are popular linear sweep generator circuits because they are efficient.

Problem 10-2. A sawtooth waveform is generated by charging and discharging a capacitor through a resistor.

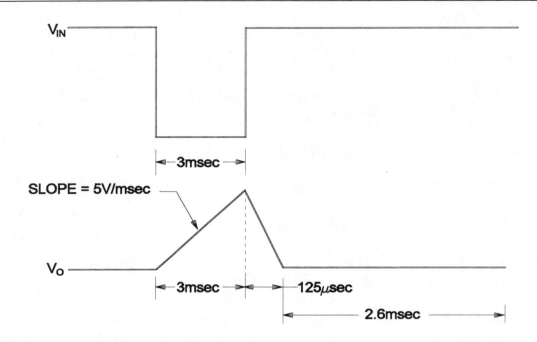

Figure P10-1. *Waveforms for Problem 10-5.*

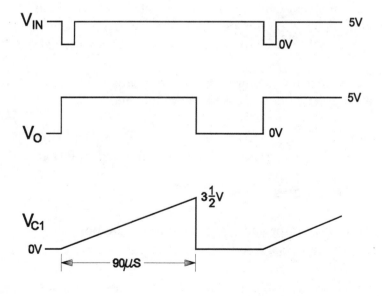

Figure P10-2. *Waveforms for Problem 10-6.*

Problem 10-3. A sawtooth waveform can be linearized by charging a capacitor with a constant current source.

Problem 10-4. $I_E = V_p/R_E = 5/100 = 50$ mA.

Problem 10-5. The waveforms are shown in *Figure P10-1*.

$t_p(max) = R1C1 = 15000 \times 0.0000002 = 3$ mS.
Output slope $= Vcc/R1C1 = 15/0.003 = 5$ V/mS.
$I_{R1} = Vcc/R1 = 15/15000 = 1$ mA.
$I_{C2} = h_{FE}I_B = h_{FE}Vcc/R3 = 50 \times 15/30,000 = 750/30,000 = 25$ mA.
$I_{disch} = I_{C2} - I_{R1} = 25 - 1 = 24$ mA.
$I_{R2} = ABS(-Vcc/R2) = Vcc/R2 = 12/10,000 = 1.2$ mA.
$t_{ret} = I_{R1}t_p(max)/I_{disch} = 0.001 \times 0.003/0.024 = 0.001/8 = 125$ uS.
$t_{rec} = I_{R1}(t_p[max] + t_{ret})/I_{R2} = 0.001(3 + 0.125) \times 0.001/0.0012$.
$t_{rec} = 0.000003125/0.0012 = 2.6$ mS.

Problem 10-6. The waveforms are shown in *Figure P10-2*.

$t = 0.667VccR3(R1 + R2)C1/[R1Vcc - V_{BE}(R1 + R2)]$.
$t = 0.667 \times 5 \times 2700(47k + 100k)0.00000001/[47,000 \times 5 - 0.6(47k + 100k)]$.
$t = 3.333 \times 2700 \times 147,000 \times 0.00000001/[235,000 - 0.6 \times 147,000]$.
$t = 13.23/[235,000 - 88,200] = 13.23/146,800 = 90.1$ uS.

Glossary

Amplifier: A device that provides a gain of more than unity.

Amplitude Modulation: The amplitude of the carrier is varied in accordance with the signal information.

Astable Multivibrator: A two-stage oscillator circuit that continuously switches between its two states.

Beta: The DC gain of a transistor.

Bistable Multivibrator: A two-stage oscillator circuit where the output is fed back to the input. It has two stable states. It is also called a flip-flop.

Blocking Oscillator: A modified Armstrong oscillator which uses an air-core transformer with a coefficient of coupling of unity.

Capacitor: An electronic component used to store an electric charge with a potential difference between its two terminals.

Choke: An electronic component that resists any change to current. Also called an inductor. Also known as a coil.

Close-Loop Gain: The gain of an amplifier with a feedback loop.

Cycle: One complete excursion of the instantaneous value of the induced emf. The portion of a periodic waveform between two adjacent corresponding points of the waveform.

Degenerative Feedback: Feedback that reduces the gain of the circuit. Also called negative feedback.

DIAC: A bidirectional dipole thyristor. It is a four-layer diode.

Diode: An electronic component that passes current when it is forward biased and blocks current when it is reverse biased.

Duty Cycle: Amount of time during which a device or signal is active.

Farad: The SI unit of capacitance.

Feedback, Negative: The output is fed back to the input 180 degrees out of phase with the input, reducing the overall gain of the circuit.

Feedback, Positive: The output is fed back to the input in phase with the input, increasing the overall gain of the circuit.

Flip-Flop: A bistable multivibrator. A one-bit memory.

Frequency: The number of cycles completed in one second.

Frequency Modulation: The frequency of the carrier is varied in accordance with the signal information.

Harmonic Frequency: An integral multiple of the fundamental frequency.

Hartley Oscillator: An oscillator circuit using a tapped coil in its tank circuit.

Henry: The SI unit of inductance. One Henry is when current changing at the rate of one ampere per second induces a counter emf of one volt into that circuit.

Inductance: The opposition to any change in the flow of electric current.

Inductive Kickback: Large counter emf generated across an inductance when an existing current is interrupted.

Inverting Amplifier: An amplifier whose output is 180 degrees out of phase with its input.

Junction, PN: Semiconductor junction that permits current to flow in one direction and blocks flow in the other direction.

Ladder Network: A network whose structure resembles a ladder.

Light-Emitting Diode: A diode that emits light when it is forward biased.

Monostable Multivibrator: An oscillator with two stages that provides positive feedback. It has only one stable state.

Multiple Tap Transformer: Transformer with more than one secondary winding.

Mutual Inductance: Equivalent coupling of a pair of magnetically coupled coils. The mutual inductance is one henry when current changing at the rate of one ampere per second in one coil induces an average emf of one volt in the other coil.

Natural Frequency: The frequency at which a closed-loop control system oscillates in response to transient changes on its inputs.

Non-Inverting Amplifier: An amplifier whose output signal is in phase with its input signal.

Open-Loop Gain: The gain of an amplifier without a feedback loop.

Operational Amplifier: An integrated circuit amplifier with high gain, high input impedance and low output impedance.

Oscillator: A circuit that generates a periodic waveform.

Period: The time taken to complete one cycle of a waveform. The reciprocal of frequency.

Periodic: A waveform that repeats itself.

Piezoelectric Effect: Generating an emf by placing a mechanical stress on a Rochelle salt crystal and displacing valence electrons within the crystal such that a surplus of electrons are at one side of the crystal and an equivalent deficiency at the other side of the crystal.

Potentiometer: A variable resistor.

Primary: Input winding of a transformer.

Q Factor: The ratio between the reactive power of either the inductance or capacitance at resonance and the real power of a resonant circuit.

Rectifier: A device that conducts current in one direction and blocks the flow of current in the other direction.

Regenerative Feedback: Positive feedback.

Resonant Circuit: Circuit where inductive and capacitive reactances are equal.

Resonant Frequency: Frequency where inductive and capacitive reactances of a circuit are equal.

Rheostat: A variable resistor.

Secondary: Output winding of a transformer.

Silicon Controlled Rectifier: A device fabricated from four layers of semiconductors. The SCR fires when an adequate current is applied to its gate. The SCR continues to conduct after the gate current is removed.

Sinusoidal: Waveform having the shape of a sinewave.

Solar Cell: A device that transforms light energy into electric energy.

Step-Down Transformer: Transformer with fewer turns on the secondary than on the primary.

Step-Up Transformer: Transformer with more turns on the secondary than on the primary.

Tank Circuit: A capacitor connected in parallel with an inductor.

Tank Current: The circulating current in a resonant tank circuit.

Threshold Voltage: The voltage level at which a device alters its response to an input signal.

Thyristor: A semiconductor device having two stable states of operation. One stable state has very low current while the other stable state has a high current that is usually limited by the resistance of the external circuit.

Time Constant: The time required for the transient current and voltages in a series RC circuit to reach 63.2 percent of their final values. t = RC

Time Constant of RL Circuit: The time it takes the current to rise to its steady-state value if it were to continue to rise at its initial rate of change for the complete time interval. The instantaneous current will reach the steady-state value of E/R after a time interval equal to five time constants has elapsed.

Toroid: A ring-shaped core.

Transistor: An active device fabricated from doped semiconductors.

TRIAC: A bidirectional triode thyristor developed to extend the negative or positive supply of an SCR to allow firing on either polarity with either positive or negative gate current pulses.

Turns Ratio: Ratio of the number of turns on the secondary of a transformer to the number of turns on the primary of the transformer.

Unijunction Transistor: A transistor fabricated with a small rod of p-material extending into the block of n-material which serves as a PN junction.

Zener Diode: A diode that has the property of avalanche breakdown when it is reverse-biased. Also called breakdown diode or reference diode.

Bibliography

Basic Electronics, Bureau of Naval Personnel Navy Training Course, NAVPERS 10087-B, Washington, D.C., 1968.

Lurch, N.E. *Fundamentals of Electronics*, 2nd Ed., John Wiley & Sons, Inc., New York, 1971.

Melen, R. and Garland, H. *Understanding IC Operational Amplifiers*, Howard W. Sams & Co., Inc., Indianapolis, 1973.

Richards, C.W. "A Tutorial on Crystal Specifications and Pullability," *Oscillator Design Handbook* (a collection from RF design), Cardiff Publishing Company, Englewood, Colorado, 1991.

RCA Solid-State Hobby Circuits Manual, RCA Corporation, Harrison, N.J., 1970.

Sedra, A.S. and Smith, K.C. *Micro-Electronic Circuits*, Holt, Rinehart and Winston, New York, 1982.

INDEX

RECTIFIER (SCR) 126, 234
SILICON CONTROLLED SWITCH 131
SILICON DEVICE 75
SILICON TRANSISTOR 68
SILICON TRANSISTORS 67
SILICON UNILATERAL SWITCH (SUS) 117, 122
SILICON UNILATERAL SWITCH DEVICES 124
SILICON-CONTROLLED RECTIFIER (SCR) 118
SILICON-CONTROLLED SWITCH (SCS) 118
SINE 153, 158
SINEWAVE 41, 42, 158
SINEWAVE GENERATOR 25
SINEWAVE OUTPUT 35, 36
SINEWAVE SHAPER 25
SINEWAVES 5
SINK 71
SINUSOIDAL 234
SINUSOIDAL OSCILLATION 21, 27, 36
SINUSOIDAL OSCILLATOR 19, 25, 26, 47, 55
SINUSOIDAL OUTPUT 27, 32, 41
SINUSOIDAL WAVEFORM 26, 38
SIREN 185
SLOPE 76
SN7401 98
SN74121 93
SN74122 93
SN74123 93
SN7413 92
SOLAR CELL VOLTAGE 181
SOLAR CELLS 181, 183, 234
SOLAR GENERATOR 181
SOLAR POWERED GENERATOR 181, 183
SOLDER BRIDGES 195, 211
SOLDER JOINTS 195, 211
SOUND EFFECTS 198

SOURCE CURRENT 15
SOURCE VOLTAGE 13
SPECIFICATIONS 28
SPEED OF LIGHT 119
SPEED-UP CAPACITORS 77
SPURIOUS OSCILLATIONS 48
SPURIOUS SIGNALS 27
SQUARE 153
SQUARE WAVEFORM 85
SQUARER 55, 61
SQUAREWAVE 5, 41, 65, 85, 158, 165, 183, 187
SQUAREWAVE GENERATOR 153
SQUAREWAVE OSCILLATOR 65
STABILITY 20
STABLE 22, 79
STABLE GAIN 9
STABLE STATE 80, 84
STABLE VALUE 59
STARTING POWER 5
STATE 65
STEADY-STATE VALUE 107
STEP VOLTAGE 75
STEP-DOWN TRANSFORMER 234
STEP-UP TRANSFORMER 106, 234
STEREOS 199
STORAGE TIME 74
STRAY CAPACITANCE 40
SURFACE 39
SURGE 38
SUS 122, 123
SUSTAINED OSCILLATION 28, 30, 35
SWEEP 144
SWEEP TRIGGER PULSE 61
SWEEP VOLTAGE 143
SWITCH 66
SWITCH TRANSISTOR 79
SWITCHES 169, 198
SWITCHING CIRCUIT 141
SWITCHING DEVICES 68

SWITCHING SPEED 73, 77, 81, 119
SWITCHING TRANSISTOR 68
SYMMETRICAL FEEDBACK LIMITER 49
SYMMETRICAL OUTPUT WAVEFORM 50
SYMMETRICAL TRANSISTOR 124
SYMMETRY 61
SYNCHRONIZED RECURRENT SWITCHING 141
SYNCHRONIZER 165

T

TANK CIRCUIT 37, 41, 47, 48, 234
TANK COIL 38
TANK CURRENT 234
TAP 34, 36, 38
TAPPED COIL 35
TAPS 36
TELEPHONE REPEATERS 9
TELESCOPING FM ANTENNA 210
TELEVISION SETS 181
TELEVISIONS 199
TEMPERATURE 27
TEMPERATURE CHANGES 48
TEMPERATURE COEFFICIENTS 47
TEMPERATURE VARIATIONS 9
TEMPERATURE-COMPENSATING CAPACITORS 47
TERMINALS 38, 123
TERTIARY LOAD WINDING 110
TEST 158
TESTING 169, 183, 210
THERMAL NOISE 27
THETA 21

PROMPT
P U B L I C A T I O N S

Digital Electronics
Stephen Kamichik

Although the field of digital electronics emerged years ago, there has never been a definitive guide to its theories, principles, and practices — until now. *Digital Electronics* is written as a textbook for a first course in digital electronics, but its applications are varied.

Useful as a guide for independent study, the book also serves as a review for practicing technicians and engineers. And because *Digital Electronics* does not assume prior knowledge of the field, the hobbyist can gain insight about digital electronics.

Some of the topics covered include analog circuits, logic gates, flip-flops, and counters. In addition, a problem set appears at the end of each chapter to test the reader's understanding and comprehension of the materials presented. Detailed instructions are provided so that the readers can build the circuits described in this book to verify their operation.

Electronic Theory
150 pages ◆ Paperback ◆ 7-3/8 x 9-1/4"
ISBN: 0-7906-1075-2 ◆ Sams: 61075
$16.95 ($22.99 Canada) ◆ February 1996

The Component Identifier and Source Book
Victor Meeldijk

Because interface designs are often reverse engineered using component data or block diagrams that list only part numbers, technicians are often forced to search for replacement parts armed only with manufacturer logos and part numbers.

This source book was written to assist technicians and system designers in identifying components from prefixes and logos, as well as find sources for various types of microcircuits and other components. There is not another book on the market that lists as many manufacturers of such diverse electronic components.

Professional Reference
384 pages ◆ Paperback ◆ 8-1/2 x 11"
ISBN: 0-7906-1088-4 ◆ Sams: 61088
$24.95 ($33.95 Canada) ◆ November 1996

CALL 1-800-428-7267 TODAY FOR THE NAME OF YOUR NEAREST PROMPT PUBLICATIONS DISTRIBUTOR